Glencoe McGraw-Hill

Math Connects

Course 3

Skills Practice Workbook

To the Student This *Skills Practice Workbook* gives you additional examples and problems for the concept exercises in each lesson. The exercises are designed to aid your study of mathematics by reinforcing important mathematical skills needed to succeed in the everyday world. The materials are organized by chapter and lesson, with one *Skills Practice* worksheet for every lesson in *Glencoe Math Connects, Course 3*.

Always keep your workbook handy. Along with your textbook, daily homework, and class notes, the completed *Skills Practice Workbook* can help you review for quizzes and tests.

To the Teacher These worksheets are the same as those found in the Chapter Resource Masters for *Glencoe Math Connects, Course 3*. The answers to these worksheets are available at the end of each Chapter Resource Masters booklet as well as in your Teacher Wraparound Edition interleaf pages.

The McGraw-Hill Companies

Copyright © by The McGraw-Hill Companies, Inc. All rights reserved. Except as permitted under the United States Copyright Act, no part of this publication may be reproduced or distributed in any form or by any means, or stored in a database or retrieval system, without prior written permission of the publisher.

Send all inquiries to:
Glencoe/McGraw-Hill
8787 Orion Place
Columbus, OH 43240

ISBN: 978-0-07-881075-6
MHID : 0-07-881075-2

Skills Practice Workbook, Course 3

Printed in the United States of America
17 18 19 20 21 QTN 19 18 17 16

CONTENTS

1-1 Skills Practice

A Plan for Problem Solving

Lesson 1-1

Use the four-step plan to solve each problem.

1. **GAS MILEAGE** Each day Ernesto drives 52 miles. If he can drive 26 miles on one gallon of gasoline, how many days can he drive on 14 gallons of gasoline?

2. **FIELD TRIP** A school policy requires that there be at least one chaperone for every 8 students on a field trip. How many chaperones are required for a field trip with 67 students?

3. **EXERCISE** Trevor jogs once every 3 days and swims once every 4 days. How often does he jog and swim on the same day?

4. **PRODUCE** At the local grocery store, lemons are 52 cents each and limes are 21 cents each. How many lemons and limes can you buy for exactly $3.75?

5. **PIZZA** The Chess Club sold 2,116 pizzas during a fundraiser that lasted for all of March, April, and May. How many pizzas did they sell per day?

6. **GUPPIES** In January, Tate's fish tank had 12 guppies. In February, it had 18, and in March it had 24. How many guppies do you expect to be in Tate's fish tank in May?

Find a pattern in the list of numbers. Then find the next number in the list.

7. 1860, 1890, 1920, 1950, 1980

8. 1024, 256, 64, 16, 4

Draw the next two figures in each of the patterns below.

9.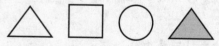

10.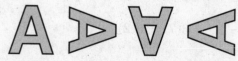

1-2 Skills Practice

Variables, Expressions, and Properties

Evaluate each expression.

1. $10 \div 2 + 8$

2. $4(9) - 36 \div 3$

3. $24 - 12 \div 4$

4. $25 + 2 \cdot 8 \div 4$

5. $49 - (3^2 + 8 \cdot 3)$

6. $2(20 - 5) + \dfrac{34 - 14}{4}$

7. $(27 + 24)(27 - 24)$

8. $2^3 \div 4 + 3 \times 6$

9. $(4 + 4) \cdot 4 + 4 \div 4$

10. $3[(8 - 2) - 5] + 7$

11. $\dfrac{28 - 7}{4^2 - 13}$

12. $(15 - 9)^2 \div (5 + 4)$

Evaluate each expression if $n = 4$, $p = 3$, and $t = 6$.

13. $3n + p$

14. $t - 2p$

15. $3p - n + 4$

16. $(np)^2$

17. np^2

18. $5(2t - n)$

19. $p(n + t)$

20. $6t^2 - t$

21. $\dfrac{npt}{3}$

22. $4(pt - 3) \div n$

23. $\dfrac{p^2 + 4}{3t - 5}$

24. $\dfrac{pn^2}{t + 10}$

25. $n^2 - 3n + 8$

26. $2t^2 - t + 9$

Name the property shown by each statement.

27. $(4 + 5)3 = 4(3) + 5(3)$

28. $1 \cdot x^2 = x^2$

29. $2(bc) = (2b)c$

30. $(6 + 2) + 5 = 6 + (2 + 5)$

31. $2(bc) = 2(cb)$

32. $(4 + 5) + 0 = 4 + 5$

33. $13 + (5 + 10) = (5 + 10) + 13$

34. $3(7 - 2) = 3(7) - 3(2)$

1-3 Skills Practice

Integers and Absolute Value

Write an integer for each situation.

1. 3 strokes below par

2. 10 strokes above par

3. a 6-yard loss

4. an 8-yard gain

5. 12 centimeters longer

6. 7 inches below normal

7. $5 off the original price

8. a gain of 6 hours

9. 2° above zero

10. a loss of 15 pounds

11. a $35 withdrawal

12. a $75 deposit

13. 1 mile above sea level

14. 20 fathoms below the surface

Replace each ● with <, >, or = to make a true sentence.

15. -12 ● 4

16. -4 ● -5

17. -10 ● -8

18. 3 ● -13

19. $|-6|$ ● $|6|$

20. $|-4|$ ● $|-5|$

Order each set of integers in each set from least to greatest.

21. $\{0, -6, 7, 2, -4\}$

22. $\{-1, -2, -3, 3, 2, 1\}$

Evaluate each expression.

23. $|-8|$

24. $|31|$

25. $|-1|$

26. $-|-256|$

27. $|3| + |-19|$

28. $|-12| + |-13|$

29. $|28| - |-26|$

30. $|28| + |-26|$

31. $|24| - |-15|$

Evaluate each expression if $a = 3$, $b = 8$, and $c = -5$.

32. $|a| + 5$

33. $|b| - 2$

34. $2|c| + b$

35. $a + |a|$

36. $|3b|$

37. $|a + 16|$

1-4 Skills Practice

Adding Integers

Add.

1. $-2 + (-3)$

2. $4 + 7$

3. $-8 + 9$

4. $12 + (-3)$

5. $-27 + 18$

6. $-11 + (-13)$

7. $-44 + 26$

8. $44 + (-26)$

9. $-15 + (-51)$

10. $(-17) + (-13)$

11. $53 + (-28)$

12. $-86 + 77$

13. $10 + (-4) + 6$

14. $-16 + (-5) + 12$

15. $-2 + 17 + (-12)$

16. $-35 + (-31) + (-39)$

17. $8 + (-12) + 15 + (-13)$

18. $-23 + (-18) + 41 + (-17)$

Evaluate each expression if $a = -9$, $b = -12$, and $c = 8$.

19. $3 + a$

20. $b + 8$

21. $-6 + c$

22. $|a| + b$

23. $|a| + |c|$

24. $|b + c|$

1-5 Skills Practice

Subtracting Integers

Subtract.

1. $6 - 7$

2. $12 - 8$

3. $-9 - 9$

4. $-17 - 18$

5. $-13 - (-25)$

6. $14 - (-19)$

7. $-25 - 15$

8. $21 - (-23)$

9. $-34 - (-11)$

10. $56 - 94$

11. $38 - (-39)$

12. $72 - 27$

13. $-36 - 47$

14. $-33 - (-68)$

15. $76 - 18$

16. $4 - |-6|$

17. $|-10| - |7|$

18. $|-52| - 49$

Evaluate each expression if $k = 8$, $m = -7$, and $p = -10$.

19. $k - 19$

20. $19 - m$

21. $p - 11$

22. $k - m$

23. $p - m$

24. $m - 3$

25. $m - k$

26. $k - m + 16$

27. $k - m - p$

Lesson 1-5

1-6 Skills Practice

Multiplying and Dividing Integers

Multiply.

1. $-2 \cdot 3$

2. $3(-3)$

3. $-4(-2)$

4. $5 \cdot 7$

5. $-9(-8)$

6. $-11 \cdot 12$

7. $15(-3)$

8. $-7(-13)$

9. $-5(2)(-7)$

10. $(-10)^2$

11. $6(8)(-3)$

12. $(-4)^3$

13. $(-9)^2$

14. $-1(-3)(-4)$

15. $(-10)^3$

16. $-3(-4)(-7)$

Divide.

17. $-15 \div 3$

18. $40 \div (-5)$

19. $-63 \div (-7)$

20. $76 \div 4$

21. $\dfrac{-56}{-4}$

22. $\dfrac{-48}{16}$

23. $\dfrac{-57}{-19}$

24. $\dfrac{75}{-5}$

Evaluate each expression if $a = -2$, $b = 5$, and $c = -6$.

25. abc

26. $2b + c$

27. $\dfrac{2b - c}{a}$

28. $ab - c$

29. $\dfrac{c}{a + b}$

30. $\dfrac{2a + c}{b}$

31. $b^2 - 5a$

32. $(-c)^2$

1-7 Skills Practice

Writing Equations

Write each verbal phrase as an algebraic expression.

1. a number divided by 5

2. the sum of d and 7

3. the product of 10 and c

4. the difference of t and 1

5. the score increased by 8 points

6. the cost split among 4 people

7. the cost of 7 CDs at $$d$ each

8. the height decreased by 2 inches

9. $500 less than the sticker price

10. the total of Ben's score and 75

11. 2 hours more than the estimate

12. 25 times the number of students

Write each verbal sentence as an algebraic equation.

13. The sum of a number and 16 is equal to 45.

14. The product of 6 and m is 216.

15. The difference of 100 and x is 57.

16. The quotient of z and 10 is equal to 32.

17. $12 less than the original price is $48.

18. 17 more than some number is equal to 85.

19. The number of members divided by 6 is 15.

20. The total of Joshua's savings and $350 is $925.

21. -65 is 5 times a number.

22. The total area decreased by 75 square feet is 250 square feet.

23. The cost of 10 books at $$d$ each is $159.50.

24. Carla's height plus 4 inches is 68 inches.

Lesson 1-7

1-8 Skills Practice

Problem-Solving Investigation: Work Backward

Use the work backward strategy to solve each problem.

1. **SKATEBOARDS** On Monday, David's skateboard shop received its first shipment of skateboards. David sold 12 skateboards that day. On Thursday, he sold 9 skateboards. On Friday, he received a shipment of 30 more skateboards and sold 10 skateboards. He then had a total of 32 skateboards in his shop. How many skateboards were delivered on Monday?

2. **SHIPPING** An overseas cargo ship was being loaded. At the end of each day, a scale showed the total weight of the ship's cargo. On Monday, 48 tons of cargo were loaded onto the ship. On Tuesday, three times as much cargo was loaded on to the ship as on Monday. On Wednesday, 68 tons of cargo were loaded onto the ship. On Thursday, 0.75 as much cargo was loaded onto the ship as on Wednesday. On Friday, 120 tons of cargo were loaded onto the ship. At the end of the day on Friday, the scale showed that the ship was carrying 690 tons of cargo. How much cargo was the ship carrying when it first came into port on Monday?

3. **NUMBERS** Jana is thinking of a number. If she divides her number by 12 and then multiplies the quotient by 8, the result is 520. What number is Jana thinking of?

4. **JOGGING** Edmund is training for a marathon. He ran a certain number of miles on Monday. On Wednesday, he ran 2 more miles than on Monday. On Saturday, he ran twice as far as on Wednesday. On Sunday, he ran 6 miles less than on Saturday. He ran 8 miles on Sunday. How many miles did Edmund run on Monday?

Use the table to solve each problem.

Airline Schedule Minneapolis, MN to Dallas, TX		
Flight Number	**Departure Time**	**Arrival Time**
253	8:20 A.M.	10:37 A.M.
142	11:52 A.M.	1:45 P.M.
295	12:00 P.M.	3:30 P.M.

5. Charles needs to take Flight 295. He needs 45 minutes to eat breakfast and pack. It takes 25 minutes to get to the airport. To be at the airport 90 minutes early, what is the latest time he can start eating breakfast?

6. Mrs. Gonzales left her office at 7:25 a.m. She planned that it would take her 30 minutes to get to the airport, but the traffic was so heavy it took an additional 20 minutes. It takes 30 minutes to check her baggage and walk to the boarding gate. What is the first flight she can take to Dallas?

1-9 Skills Practice

Solving Addition and Subtraction Equations

Solve each equation. Check your solution.

1. $x + 3 = 4$

2. $y + 6 = 5$

3. $t - 2 = 2$

4. $z - 5 = 1$

5. $a + 4 = -3$

6. $h - 3 = -6$

7. $u - 4 = -1$

8. $8 + d = 14$

9. $19 = x + 7$

10. $17 = b - 8$

11. $-19 = z - 21$

12. $22 = y + 29$

13. $16 = 24 + p$

14. $-17 = 19 + x$

15. $f - 25 = 35$

16. $y + 37 = 59$

17. $s + 46 = 72$

18. $m + 65 = 11$

19. $r + 53 = -19$

20. $n - 75 = 42$

21. $g - 35 = -62$

22. $111 = x + 68$

23. $-54 = -32 + w$

24. $-27 + z = 47$

Lesson 1-9

1-10 Skills Practice

Solving Multiplication and Division Equations

Solve each equation. Check your solution.

1. $\dfrac{u}{7} = 3$

2. $3c = 12$

3. $5x = -15$

4. $-7z = 49$

5. $\dfrac{n}{3} = -7$

6. $\dfrac{a}{-9} = -11$

7. $-14g = -56$

8. $\dfrac{t}{-12} = 11$

9. $18y = -144$

10. $135 = 9z$

11. $11d = -143$

12. $116 = -29k$

13. $\dfrac{w}{9} = 17$

14. $-14 = \dfrac{y}{-7}$

15. $-112 = -8v$

16. $17c = 136$

17. $-21a = -126$

18. $\dfrac{s}{-19} = 9$

19. $\dfrac{m}{-31} = -7$

20. $16q = 272$

21. $15 = \dfrac{z}{-14}$

22. $\dfrac{g}{-22} = -23$

23. $\dfrac{y}{25} = 16$

24. $47k = 517$

2-1 Skills Practice

Rational Numbers

Write each fraction or mixed number as a decimal.

1. $\dfrac{1}{10}$

2. $\dfrac{1}{8}$

3. $-\dfrac{3}{4}$

4. $-\dfrac{4}{5}$

5. $\dfrac{21}{50}$

6. $-3\dfrac{9}{20}$

7. $4\dfrac{9}{25}$

8. $\dfrac{7}{9}$

9. $1\dfrac{1}{6}$

10. $-2\dfrac{4}{15}$

11. $\dfrac{5}{33}$

12. $7\dfrac{3}{11}$

Write each decimal as a fraction or mixed number in simplest form.

13. 0.9

14. 0.7

15. 0.84

16. 0.92

17. −1.12

18. −5.05

19. 2.35

20. 8.85

21. $-0.\overline{1}$

22. $4.\overline{8}$

23. $6.\overline{7}$

24. $-8.\overline{4}$

Lesson 2-1

2-2 Skills Practice

Comparing and Ordering Rational Numbers

Replace each ● with <, >, or = to make a true sentence.

1. $\frac{1}{2}$ ● $\frac{3}{4}$

2. $\frac{1}{3}$ ● $\frac{1}{6}$

3. $\frac{2}{5}$ ● $\frac{3}{10}$

4. $\frac{2}{9}$ ● $\frac{1}{3}$

5. $\frac{3}{4}$ ● $\frac{9}{12}$

6. $\frac{3}{8}$ ● $\frac{2}{5}$

7. $-\frac{5}{6}$ ● $-\frac{6}{7}$

8. $-\frac{4}{9}$ ● $-\frac{5}{11}$

9. $\frac{5}{9}$ ● 0.55

10. 4.72 ● $4\frac{10}{13}$

11. $-2\frac{7}{15}$ ● -2.45

12. 5.25 ● $5.\overline{25}$

13. -1.62 ● $-1\frac{5}{8}$

14. $11\frac{4}{9}$ ● $11.\overline{4}$

15. $-1.\overline{27}$ ● $-1.2\overline{7}$

Order each set of rational numbers from least to greatest.

16. $0.3, 0.2, \frac{1}{3}, \frac{2}{9}$

17. $1\frac{2}{5}, 1\frac{2}{3}, 1.55, 1.67$

18. $2.7, 2\frac{1}{7}, 3.13, 1\frac{9}{10}$

19. $\frac{1}{4}, -1.7, 0.2, -1\frac{3}{4}$

20. $-2.21, -2.09, -2\frac{1}{9}, -1\frac{10}{11}$

21. $-3.1, 2.75, 1\frac{7}{8}, -\frac{2}{3}$

22. $6\frac{7}{8}, 6\frac{15}{16}, 6.9, 5.3$

23. $-4\frac{1}{6}, -4.19, -5.3, -5\frac{1}{3}$

24. $5\frac{9}{11}, 5.93, 5\frac{7}{20}, 5.81$

25. $-3\frac{1}{4}, -4\frac{1}{8}, -3.65, -3\frac{4}{11}, -4.05$

2-3 **Skills Practice**

Multiplying Positive and Negative Fractions

Multiply. Write in simplest form.

1. $\frac{1}{8} \cdot \frac{2}{3}$

2. $\frac{2}{9} \cdot \frac{7}{8}$

3. $\frac{5}{6} \cdot \frac{3}{11}$

4. $-\frac{4}{7} \cdot \frac{3}{10}$

5. $\frac{2}{9} \cdot \left(-\frac{3}{8}\right)$

6. $-\frac{3}{5} \cdot \left(-\frac{5}{9}\right)$

7. $1\frac{3}{4} \cdot \frac{2}{3}$

8. $\frac{4}{5} \cdot 4\frac{3}{8}$

9. $-\frac{2}{15} \cdot 5\frac{5}{6}$

10. $-1\frac{3}{7} \cdot 1\frac{1}{5}$

11. $-2\frac{1}{4} \cdot 1\frac{2}{3}$

12. $1\frac{9}{16} \cdot 2\frac{4}{5}$

13. $-3\frac{1}{7} \cdot \left(-1\frac{2}{11}\right)$

14. $2\frac{2}{3} \cdot \left(-2\frac{1}{4}\right)$

15. $\left(-\frac{4}{5}\right)\left(-\frac{4}{5}\right)$

ALGEBRA Evaluate each expression if $r = \frac{5}{6}$, $s = -\frac{1}{3}$, $t = \frac{4}{5}$, **and** $v = -\frac{3}{4}$.

16. rv

17. st

18. rs

19. stv

20. rst

21. rtv

ALGEBRA Evaluate each expression if $a = -\frac{5}{9}$, $b = -\frac{1}{5}$, $c = \frac{2}{3}$, **and** $d = \frac{3}{4}$.

22. ad

23. bc

24. abc

2-4 Skills Practice

Dividing Positive and Negative Fractions

Write the multiplicative inverse of each number.

1. $\dfrac{2}{3}$

2. $-\dfrac{4}{7}$

3. $-\dfrac{1}{12}$

4. 22

5. $\dfrac{9}{35}$

6. $-\dfrac{14}{17}$

7. $1\dfrac{5}{7}$

8. $-1\dfrac{3}{13}$

9. $2\dfrac{3}{7}$

10. $-3\dfrac{6}{11}$

11. $4\dfrac{8}{15}$

12. $5\dfrac{3}{5}$

Divide. Write in simplest form.

13. $\dfrac{3}{7} \div \dfrac{3}{5}$

14. $\dfrac{2}{7} \div \dfrac{6}{7}$

15. $-\dfrac{5}{8} \div \dfrac{3}{4}$

16. $\dfrac{7}{9} \div \left(-\dfrac{14}{15}\right)$

17. $-\dfrac{4}{5} \div \dfrac{8}{9}$

18. $\dfrac{2}{11} \div \dfrac{4}{9}$

19. $1\dfrac{3}{4} \div 2\dfrac{1}{2}$

20. $-2\dfrac{3}{5} \div 1\dfrac{3}{10}$

21. $-3\dfrac{4}{7} \div \left(-1\dfrac{1}{14}\right)$

22. $\dfrac{10}{11} \div 5$

23. $-4 \div \dfrac{3}{5}$

24. $3\dfrac{4}{15} \div 4\dfrac{2}{3}$

25. $9\dfrac{1}{3} \div 5\dfrac{3}{5}$

26. $-12\dfrac{3}{4} \div \left(-2\dfrac{5}{6}\right)$

27. $2\dfrac{4}{9} \div \left(-6\dfrac{2}{7}\right)$

28. $-11\dfrac{1}{5} \div 3\dfrac{1}{9}$

2-5 Skills Practice

Adding and Subtracting Like Fractions

Add or subtract. Write in simplest form.

1. $\dfrac{1}{5} + \dfrac{3}{5}$

2. $\dfrac{2}{9} + \dfrac{5}{9}$

3. $\dfrac{7}{11} + \dfrac{3}{11}$

4. $-\dfrac{1}{4} + \dfrac{3}{4}$

5. $-\dfrac{4}{9} + \dfrac{8}{9}$

6. $-\dfrac{5}{7} + \dfrac{2}{7}$

7. $\dfrac{7}{12} + \dfrac{5}{12}$

8. $\dfrac{1}{9} + \left(-\dfrac{4}{9}\right)$

9. $-\dfrac{5}{7} + \left(-\dfrac{3}{7}\right)$

10. $-\dfrac{9}{16} + \left(-\dfrac{3}{16}\right)$

11. $\dfrac{5}{8} - \dfrac{3}{8}$

12. $\dfrac{13}{19} - \dfrac{6}{19}$

13. $\dfrac{2}{7} - \dfrac{6}{7}$

14. $\dfrac{4}{15} - \dfrac{7}{15}$

15. $\dfrac{1}{9} - \left(-\dfrac{4}{9}\right)$

16. $\dfrac{3}{13} - \left(-\dfrac{11}{13}\right)$

17. $2\dfrac{3}{7} + 1\dfrac{2}{7}$

18. $1\dfrac{4}{15} + 4\dfrac{8}{15}$

19. $5\dfrac{6}{7} - 3\dfrac{2}{7}$

20. $6\dfrac{7}{12} - 3\dfrac{1}{12}$

21. $-2\dfrac{5}{11} - 7\dfrac{1}{11}$

22. $-4\dfrac{3}{8} - 2\dfrac{7}{8}$

23. $5\dfrac{2}{9} - 2\dfrac{4}{9}$

24. $8\dfrac{1}{5} - 4\dfrac{2}{5}$

Lesson 2-5

2-6 Skills Practice

Adding and Subtracting Unlike Fractions

Add or subtract. Write in simplest form.

1. $\frac{1}{6} + \frac{1}{2}$

2. $\frac{4}{9} + \frac{1}{3}$

3. $\frac{7}{8} + \frac{1}{4}$

4. $\frac{3}{4} + \frac{2}{3}$

5. $\frac{6}{7} - \frac{3}{14}$

6. $\frac{4}{5} - \frac{1}{3}$

7. $\frac{1}{4} - \frac{5}{6}$

8. $-\frac{3}{5} + \frac{1}{4}$

9. $-\frac{3}{7} - \frac{2}{3}$

10. $\frac{4}{7} - \left(-\frac{1}{2}\right)$

11. $3\frac{2}{5} + 2\frac{1}{3}$

12. $5\frac{5}{7} + 3\frac{1}{2}$

13. $3\frac{1}{6} + 4\frac{1}{4}$

14. $1\frac{1}{2} + \left(-1\frac{1}{5}\right)$

15. $2\frac{3}{4} + \left(-6\frac{3}{8}\right)$

16. $5\frac{1}{4} + \left(-2\frac{2}{3}\right)$

17. $-5\frac{1}{12} - 3\frac{2}{3}$

18. $-3\frac{3}{5} - \frac{9}{10}$

19. $-2\frac{1}{5} - 3\frac{3}{4}$

20. $2\frac{1}{3} - \left(-4\frac{5}{6}\right)$

21. $3\frac{2}{7} - \left(-4\frac{2}{3}\right)$

22. $5\frac{7}{9} - \left(-2\frac{1}{3}\right)$

23. $10\frac{2}{9} - \left(-3\frac{1}{3}\right)$

24. $-2\frac{1}{3} - \left(-5\frac{4}{5}\right)$

2-7 **Skills Practice**

Solving Equations with Rational Numbers

Solve each equation. Check your solution.

1. $x + 2.62 = 6.37$

2. $y - 3.16 = 7.92$

3. $-3.38 = r - 9.76$

4. $s + \dfrac{5}{8} = \dfrac{7}{8}$

5. $-\dfrac{5}{6} = x - \dfrac{1}{3}$

6. $-\dfrac{4}{5} + z = \dfrac{1}{10}$

7. $3.4c = 6.8$

8. $-1.56 = 0.26w$

9. $12.8y = 6.4$

10. $\dfrac{3}{4}x = 9$

11. $\dfrac{4}{9} = \dfrac{8}{11}a$

12. $-\dfrac{2}{5}s = \dfrac{4}{15}$

13. $-\dfrac{2}{3} = \dfrac{3}{10}t$

14. $-\dfrac{4}{11}w = -\dfrac{19}{22}$

15. $5.1 = -1.7r$

16. $z - (-3.2) = 3.69$

17. $-2.11 = w - (-5.81)$

18. $\dfrac{w}{2.6} = 3.5$

19. $-\dfrac{x}{1.8} = 7.2$

20. $2\dfrac{1}{4}y = 3\dfrac{3}{8}$

21. $-2\dfrac{2}{5}f = -3\dfrac{1}{5}$

22. $1.5d = \dfrac{3}{8}$

23. $-7.5g = -6\dfrac{2}{3}$

24. $-2\dfrac{1}{5} = c - \left(-\dfrac{4}{5}\right)$

2-8 Skills Practice

Problem-Solving Investigation: Look for a Pattern

Look for a pattern. Then use the pattern to solve each problem.

1. **YARN** A knitting shop is having a huge yarn sale. One skein sells for $1.00, 2 skeins sell for $1.50, and 3 skeins sell for $2.00. If this pattern continues, how many skeins of yarn can you buy for $5.00?

2. **BIOLOGY** Biologists place sensors in 8 concentric circles to track the movement of grizzly bears throughout Yellowstone National Park. Four sensors are placed in the inner circle. Eight sensors are placed in the next circle. Sixteen sensors are placed in the third circle, and so on. If the pattern continues, how many sensors are needed in all?

3. **HONOR STUDENTS** A local high school displays pictures of the honor students from each school year on the office wall. The top row has 9 pictures displayed. The next 3 rows have 7, 10, and 8 pictures displayed. The pattern continues to the bottom row, which is the first row with 14 pictures in it. How many rows of pictures are there on the office wall?

4. **CHEERLEADING** The football cheerleaders will arrange themselves in rows to form a pattern on the football field at halftime. In the first five rows there are 12, 10, 11, 9, and 10 girls in each row. They will form a total of twelve rows. If the pattern continues, how many girls will be in the back row?

5. **GEOMETRY** Find the perimeters of the next two figures in the pattern. The length of each side of each small square is 3 feet.

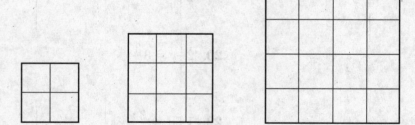

6. **HOT TUBS** A hot tub holds 630 gallons of water when it is full. A hose fills the tub at a rate of 6 gallons every five minutes. How long will it take to fill the hot tub?

2-9 Skills Practice

Powers and Exponents

Write each expression using exponents.

1. $2 \cdot 2 \cdot 2 \cdot 2$

2. $9 \cdot 9$

3. $7 \cdot 7 \cdot 7 \cdot 7 \cdot 7 \cdot 7$

4. $x \cdot x \cdot x$

5. $c \cdot c \cdot c \cdot c \cdot c$

6. $s \cdot s \cdot s \cdot s \cdot s \cdot s \cdot s$

7. $5 \cdot 5 \cdot 5 \cdot 3 \cdot 3$

8. $4 \cdot 4 \cdot 4 \cdot 4 \cdot 6 \cdot 6 \cdot 6$

9. $8 \cdot 8 \cdot 2 \cdot 2 \cdot 2 \cdot 2 \cdot 8$

10. $a \cdot a \cdot b \cdot a \cdot b \cdot a \cdot a$

11. $m \cdot n \cdot n \cdot n \cdot m \cdot n$

12. $y \cdot x \cdot x \cdot y \cdot x \cdot y \cdot y$

Evaluate each expression.

13. 4^3

14. 2^5

15. 8^3

16. 5^4

17. 2^8

18. $2^3 \cdot 5^2$

19. $4^2 \cdot 3^4$

20. $2^6 \cdot 6^2$

21. $3^3 \cdot 7^3$

22. 2^{-3}

23. 8^{-2}

24. 7^{-4}

19

Lesson 2-9

2-10 Skills Practice

Scientific Notation

Write each number in standard form.

1. 6.7×10^1

2. 6.1×10^4

3. 1.6×10^3

4. 3.46×10^2

5. 2.91×10^5

6. 8.651×10^7

7. 3.35×10^{-1}

8. 7.3×10^{-6}

9. 1.49×10^{-7}

10. 4.0027×10^{-4}

11. 5.2277×10^{-3}

12. 8.50284×10^{-2}

Write each number in scientific notation.

13. 34

14. 273

15. 79,700

16. 6,590

17. 4,733,800

18. 2,204,000,000

19. 0.00916

20. 0.29

21. 0.00000571

22. 0.0008331

23. 0.0121

24. 0.00000018

3-1 Skills Practice

Square Roots

Find each square root.

1. $\sqrt{16}$

2. $-\sqrt{9}$

3. $\sqrt{36}$

4. $\sqrt{196}$

5. $\sqrt{121}$

6. $-\sqrt{81}$

7. $-\sqrt{0.04}$

8. $\sqrt{289}$

9. $\sqrt{0.81}$

10. $-\sqrt{400}$

11. $\sqrt{\dfrac{16}{49}}$

12. $\sqrt{\dfrac{49}{100}}$

ALGEBRA Solve each equation.

13. $s^2 = 81$

14. $t^2 = 36$

15. $x^2 = 49$

16. $256 = z^2$

17. $900 = y^2$

18. $1{,}024 = h^2$

19. $c^2 = \dfrac{49}{64}$

20. $a^2 = \dfrac{25}{121}$

21. $\dfrac{1}{100} = d^2$

22. $\dfrac{144}{169} = r^2$

23. $b^2 = \dfrac{9}{441}$

24. $x^2 = \dfrac{121}{400}$

3-2 Skills Practice

Estimating Square Roots

Estimate to the nearest whole number.

1. $\sqrt{5}$

2. $\sqrt{18}$

3. $\sqrt{10}$

4. $\sqrt{34}$

5. $\sqrt{53}$

6. $\sqrt{80}$

7. $\sqrt{69}$

8. $\sqrt{99}$

9. $\sqrt{120}$

10. $\sqrt{77}$

11. $\sqrt{171}$

12. $\sqrt{230}$

13. $\sqrt{147}$

14. $\sqrt{194}$

15. $\sqrt{290}$

16. $\sqrt{440}$

17. $\sqrt{578}$

18. $\sqrt{730}$

19. $\sqrt{1,010}$

20. $\sqrt{1,230}$

21. $\sqrt{8.42}$

22. $\sqrt{17.8}$

23. $\sqrt{11.5}$

24. $\sqrt{37.7}$

25. $\sqrt{23.8}$

26. $\sqrt{59.4}$

27. $\sqrt{97.3}$

28. $\sqrt{118.4}$

29. $\sqrt{84.35}$

30. $\sqrt{45.92}$

3-3 **Skills Practice**

Problem-Solving Investigation: Use a Venn Diagram

Use a Venn diagram to solve each problem.

1. **PHONE SERVICE** Of the 5,750 residents of Homer, Alaska, 2,330 pay for landline phone service and 4,180 pay for cell phone service. One thousand seven hundred fifty pay for both landline and cell phone service. How many residents of Homer do not pay for any type of phone service?

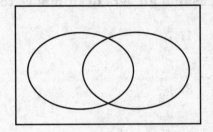

2. **BIOLOGY** Of the 2,890 ducks living in a particular wetland area, scientists find that 1,260 have deformed beaks, while 1,320 have deformed feet. Six hundred ninety of the birds have both deformed feet and beaks. How many of the ducks living in the wetland area have no deformities?

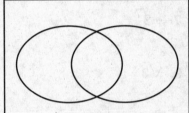

3. **FLU SYMPTOMS** The local health agency treated 890 people during the flu season. Three hundred fifty of the patients had flu symptoms, 530 had cold symptoms, and 140 had both cold and flu symptoms. How many of the patients treated by the health agency had no cold or flu symptoms?

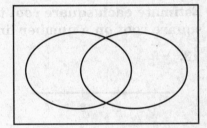

4. **HOLIDAY DECORATIONS** During the holiday season, 13 homes on a certain street displayed lights and 8 displayed lawn ornaments. Five of the homes displayed both lights and lawn ornaments. If there are 32 homes on the street, how many had no decorations at all?

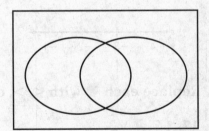

5. **LUNCHTIME** At the local high school, 240 students reported they have eaten the cafeteria's hot lunch, 135 said they have eaten the cold lunch, and 82 said they have eaten both the hot and cold lunch. If there are 418 students in the school, how many bring lunch from home?

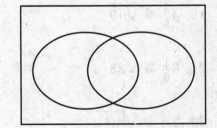

Lesson 3-3

3-4 Skills Practice

The Real Number System

Name all sets of numbers to which each real number belongs.

1. 12

2. -15

3. $1\frac{1}{2}$

4. 3.18

5. $\frac{8}{4}$

6. $9.\overline{3}$

7. $-2\frac{7}{9}$

8. $\sqrt{25}$

9. $\sqrt{3}$

10. $-\sqrt{64}$

11. $-\sqrt{12}$

12. $\sqrt{13}$

Estimate each square root to the nearest tenth. Then graph the square root on a number line.

13. $\sqrt{5}$

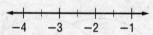

14. $\sqrt{14}$

15. $-\sqrt{6}$

16. $-\sqrt{13}$

Replace each ● with <, >, or = to make a true sentence.

17. $1.7 ● \sqrt{3}$

18. $\sqrt{6} ● 2\frac{1}{2}$

19. $4\frac{2}{5} ● \sqrt{19}$

20. $4.\overline{8} ● \sqrt{24}$

21. $6\frac{1}{6} ● \sqrt{38}$

22. $\sqrt{55} ● 7.4\overline{2}$

23. $2.1 ● \sqrt{4.41}$

24. $2.\overline{7} ● \sqrt{7.7}$

3-7 Skills Practice

Distance on the Coordinate Plane

Find the distance between each pair of points whose coordinates are given. Round to the nearest tenth if necessary.

1.

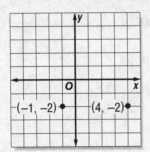

2.

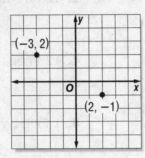

3.

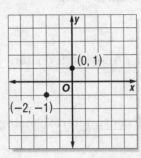

4.

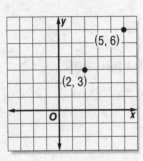

5.

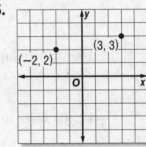

6.

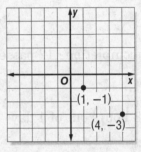

Graph each pair of ordered pairs. Then find the distance between the points. Round to the nearest tenth if necessary.

7. $(-3, 0), (3, -2)$

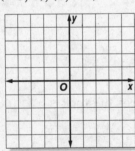

8. $(-4, -3), (2, 1)$

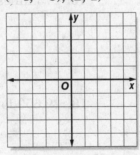

9. $(0, 2), (5, -2)$

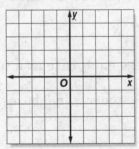

10. $(-2, 1), (-1, 2)$

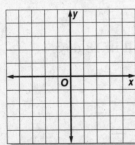

11. $(0, 0), (-4, -3)$

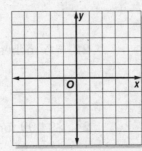

12. $(-3, 4), (2, -3)$

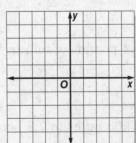

Lesson 3-7

4-1 Skills Practice

Ratios and Rates

Express each ratio in simplest form.

1. 15 cats:50 dogs

2. 18 adults to 27 teens

3. 27 nurses to 9 doctors

4. 12 losses in 32 games

5. 50 centimeters:1 meter

6. 1 foot:1 yard

7. 22 players:2 teams

8. $28:8 pounds

9. 8 completions:12 passes

10. 21 hired out of 105 applicants

11. 18 hours out of 1 day

12. 64 boys to 66 girls

13. 66 miles on 4 gallons

14. 48 wins:18 losses

15. 112 peanuts:28 cashews

16. 273 miles in 6 hours

Express each rate as a unit rate.

17. 96 students in 3 buses

18. $9,650 for 100 shares of stock

19. $21.45 for 13 gallons of gasoline

20. 125 meters in 10 seconds

21. 30.4 pounds of tofu in 8 weeks

22. 6.5 inches of rainfall in 13 days

23. 103.68 miles in 7.2 hours

24. $94.99 for 7 pizzas

4-2 Skills Practice

Proportional and Nonproportional Relationships

For Exercises 1–3, use the table of values. Write the ratios in the table to show the relationship between each set of values.

1.
Number of Hours	1	2	3	4
Total Amount Earned	$15	$30	$45	$60
Ratios				

2.
Number of Packages	1	2	3	4
Total Cost	$11	$20	$29	$38
Ratios				

3.
Number of Classrooms	1	2	3	4
Total Students	24	48	72	92
Ratios				

For Exercises 4–8 use the table of values. Write *proportional* or *nonproportional*.

4.
Number of Hours	1	2	3	4
Total Amount Earned	$0.99	$1.98	$2.97	$3.96

5.
Number of Hours	1	2	3	4
Total Amount Earned	$17.25	$35.50	$50.75	$70

6.
Number of Hours	1	2	3	4
Number of Pages Read in Book	37	73	109	145

7.
Number of Lunches	1	2	3	4
Total Cost	$2.75	$5.50	$8.25	$11

8. Jack is ordering pies for a family reunion. Each pie costs $4.50. For orders smaller than a dozen pies, there is a $5 delivery charge. Is the cost proportional to the number of pies ordered?

Lesson 4-2

4-3　Skills Practice

Rate of Change

TEMPERATURE Use the table below that shows the high temperature of a city for the first part of August.

Date	1	5	14	15
High Temperature (°F)	85	93	102	102

1. Find the rate of change in the high temperature between August 1 and August 5.

2. Find the rate of change in the high temperature between August 5 and August 14.

3. During which of these two time periods did the high temperature rise faster?

4. Find the rate of change in the high temperature between August 14 and August 15. Then interpret its meaning.

COMPANY GROWTH Use the graph that shows the number of employees at a company between 2000 and 2008.

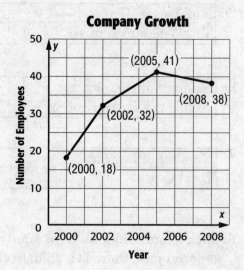

5. Find the rate of change in the number of employees between 2000 and 2002.

6. Find the rate of change in the number of employees between 2002 and 2005.

7. During which of these two time periods did the number of employees grow faster?

8. Find the rate of change in the number of employees between 2005 and 2008. Then interpret its meaning.

4-4 Skills Practice

Constant Rate of Change

Determine whether the relationship between the two quantities described in each table is linear. If so, find the constant rate of change. If not, explain your reasoning.

1. $10 per hour

Hours spent Babysitting	Money Earned
1	10
3	30
5	50
7	70

2. 4° per hour

Time	Temperature
9	60
10	64
11	68
12	72

3. No, the change in number of magazines sold is not constant

Number of Students	Number of Magazines Sold
10	100
15	110
20	200
25	240

4. No, the change in number of apples per tree is not constant

Number of Trees	Number of Apples
5	100
10	120
15	150
20	130

5. 3 hours per volunteer

Number of Volunteers	Number of Hours Logged
5	15
10	30
15	45
20	60

6. decreasing 2 gallons per 50 miles; or 1 gallon per 25 miles

Gas Left in Tank	Miles Driven
12	0
10	50
8	100
6	150

Find the constant rate of change for each graph and interpret its meaning.

9. −2° F per hour; the temperature if decreasing by 2° F per hour.

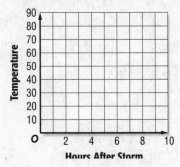

10. 3 lbs per person; Three pounds of meat are needed per person

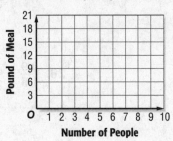

Lesson 4-4

4-5 Skills Practice

Solving Proportions

Determine whether each pair of ratios forms a proportion.

1. $\dfrac{5}{8}, \dfrac{2}{3}$

2. $\dfrac{7}{3}, \dfrac{14}{6}$

3. $\dfrac{6}{8}, \dfrac{9}{12}$

4. $\dfrac{16}{9}, \dfrac{11}{6}$

5. $\dfrac{55}{10}, \dfrac{12}{2}$

6. $\dfrac{6}{8}, \dfrac{15}{20}$

7. $\dfrac{5}{9}, \dfrac{15}{27}$

8. $\dfrac{3}{18}, \dfrac{11}{66}$

9. $\dfrac{7}{11}, \dfrac{15}{23}$

10. $\dfrac{9}{13}, \dfrac{13}{17}$

11. $\dfrac{3}{42}, \dfrac{5}{70}$

12. $\dfrac{6}{7}, \dfrac{36}{49}$

Solve each proportion.

13. $\dfrac{4}{12} = \dfrac{y}{9}$

14. $\dfrac{6}{18} = \dfrac{4}{c}$

15. $\dfrac{7}{z} = \dfrac{84}{12}$

16. $\dfrac{5}{10} = \dfrac{8}{w}$

17. $\dfrac{x}{9} = \dfrac{4}{15}$

18. $\dfrac{6}{20} = \dfrac{y}{5}$

19. $\dfrac{5}{9} = \dfrac{6}{r}$

20. $\dfrac{8}{n} = \dfrac{10}{7}$

21. $\dfrac{d}{5} = \dfrac{12}{80}$

22. $\dfrac{y}{5} = \dfrac{13}{10}$

23. $\dfrac{2}{28} = \dfrac{p}{35}$

24. $\dfrac{11}{t} = \dfrac{100}{11}$

25. $\dfrac{1.2}{m} = \dfrac{3}{5}$

26. $\dfrac{0.9}{1.5} = \dfrac{a}{10}$

27. $\dfrac{3}{7} = \dfrac{k}{4.2}$

28. $\dfrac{6.3}{x} = \dfrac{18}{5}$

29. $\dfrac{3.6}{9} = \dfrac{b}{0.5}$

30. $\dfrac{14}{1.5} = \dfrac{4.2}{y}$

4-6 Skills Practice

Problem-Solving Investigation: Draw a Diagram

For Exercises 1–5, use the draw a diagram strategy to solve the problem.

1. **AQUARIUM** An aquarium holds 60 gallons of water. After 6 minutes, the tank has 15 gallons of water in it. How many more minutes will it take to fill the tank?

2. **TILING** Meredith has a set of ninety 1-inch tiles. If she starts with one tile, then surrounds it with a ring of tiles to create a larger square, how many surrounding rings can she make before she runs out of tiles?

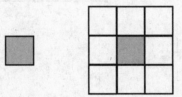

3. **SEWING** Judith has a 30-yard by 1-yard roll of fabric. She needs to use 1.5 square yards to create one costume. How many costumes can she create?

4. **DRIVING** It takes 3 gallons of gas to drive 102 miles. How many miles can be driven on 16 gallons of gas?

5. **PACKING** Hector can fit 75 compact discs into 5 boxes. How many compact discs can he fit into 14 boxes?

Lesson 4-6

4-7 Skills Practice

Similar Polygons

Determine whether each pair of polygons is similar. Explain.

1.
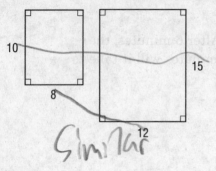
10   15
8   12

Similar

2.
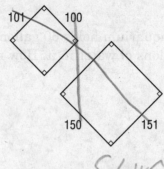
101   100
150   151

Similar
aii 70

3.
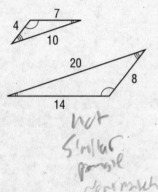
4   7
10
20
8
14

not
similar
parrell
doesn't match

4.
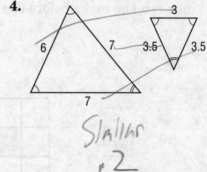
3
6   7   3.5   3.5
7

Similar
·2

Each pair of polygons is similar. Write a proportion to find each missing measure. Then solve.

5.
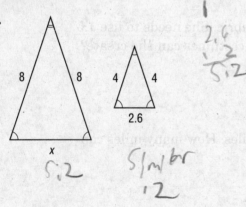
8   8
4   4
2.6
x

2.6
·2
5.2

5.2   Similar
·2

6.
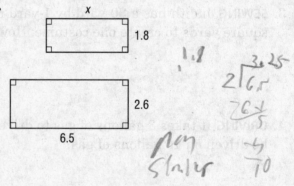
x
1.8
2.6
6.5

1.8   3.25
2⟌6.5
26↓
not   70
Similar

7.
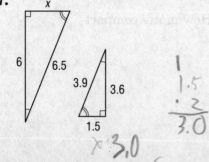
x
6   6.5
3.9   3.6
1.5

1.5
·2
3.0

x·3.0
Similar

8.
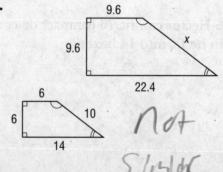
9.6
9.6   x
22.4
6
6   10
14

not
Similar

4-8 Skills Practice

Dilations

Find the coordinates of the vertices of triangle $A'B'C'$ after triangle ABC is dilated using the given scale factor. Then graph triangle ABC and its dilation.

1. $A(1, 1)$, $B(1, 3)$, $C(3, 1)$; scale factor 3

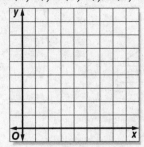

2. $A(-2, -2)$, $B(-1, 2)$, $C(2, 1)$; scale factor 2

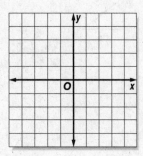

3. $A(-4, 6)$, $B(2, 6)$, $C(0, 8)$; scale factor $\frac{1}{2}$

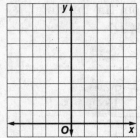

4. $A(-3, -2)$, $B(1, 2)$, $C(2, -3)$; scale factor 1.5

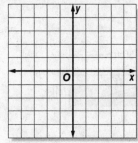

Segment $P'Q'$ is a dilation of segment PQ. Find the scale factor of the dilation and classify it as an *enlargement* or a *reduction*.

5.

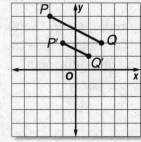

6.

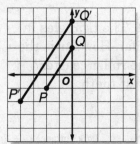

7.

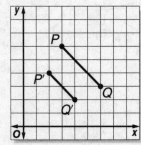

8.

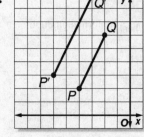

Lesson 4-8

4-9 Skills Practice

Indirect Measurement

Write a proportion and solve the problem.

1. HEIGHT How tall is Becky?

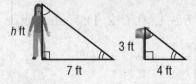

2. FLAGS How tall is the flagpole?

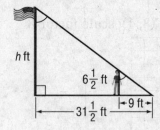

3. BEACH How deep is the water 50 feet from shore?

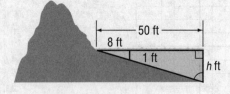

4. ACCESSIBILITY How high is the ramp when it is 2 feet from the building? (*Hint:* △ABE ~ △ACD)

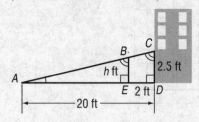

5. AMUSEMENT PARKS The triangles in the figure are similar. How far is the water ride from the roller coaster? Round to the nearest tenth.

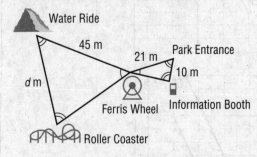

6. CLASS CHANGES The triangles in the figure are similar. How far is the entrance to the gymnasium from the band room?

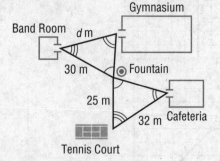

5-4 Skills Practice

Finding Percents Mentally

Compute mentally.

1. 50% of 40

2. 25% of 36

3. 10% of 60

4. 1% of 100

5. 20% of 15

6. 40% of 30

7. $33\frac{1}{3}$% of 21

8. $12\frac{1}{2}$% of 32

9. 75% of 28

10. 10% of 230

11. 90% of 30

12. $83\frac{1}{3}$% of 18

13. 1% of 300

14. $62\frac{1}{2}$% of 24

15. 60% of 45

16. 70% of 50

17. $16\frac{2}{3}$% of 48

18. 10% of 66

19. 30% of 70

20. 1% of 240

21. $66\frac{2}{3}$% of 51

22. 10% of 45

23. 1% of 73

24. 10% of 12.4

25. 1% of 18.9

26. 10% of 107

27. 1% of 153

28. $87\frac{1}{2}$% of 72

29. $83\frac{1}{3}$% of 54

30. $62\frac{1}{2}$% of 64

Lesson 5-4

5-5 Skills Practice

Problem-Solving Investigation: Reasonable Answers

For Exercises 1–12, estimate and rewrite the problem to determine a reasonable answer.

1. 53% of 813

2. 27% of 456

3. 87% of 1,978

4. 11% of 176

5. 67% of 543

6. 8% of 697

7. 81% of 2,211

8. 48% of 762

9. 4% of 4,874

10. 23% of 584

11. 45% of 1,252

12. 32% of 620

For Exercises 13–24, estimate and rewrite the problem to determine a reasonable answer.

13. $54.87 + $28.97

14. $22.38 + $46.12

15. $94.67 + $17.78

16. $88.88 + $36.32

17. $7.87 + $48.31

18. $74.78 + $75.18

19. $37.42 + $85.01

20. $28.69 + $35.09

21. $108.24 + $127.95

22. $89.99 + $79.99

23. $217.87 + $186.65

24. $46.22 + $86.86

..

5-8 Skills Practice

Percent of Change

Find each percent of change. Round to the nearest tenth of a percent if necessary. State whether the percent of change is an *increase* or a *decrease*.

1. original: 4
new: 6

2. original: 35
new: 28

3. original: 80
new: 52

4. original: 45
new: 63

5. original: 120
new: 132

6. original: 210
new: 105

7. original: 84
new: 111

8. original: 91
new: 77

Find the selling price for each item given the cost to the store and the markup.

9. suit: $200, 50% markup

10. tire: $50, 40% markup

11. sport bag: $40, 30% markup

12. radio: $120, 25% markup

13. grill: $85, 15% markup

14. microwave: $96, 20% markup

15. chair: $140, 45% markup

16. camcorder: $350, 33% markup

17. camera: $245, 10% markup

18. diamond ring: $470, 35% markup

Find the sale price of each item to the nearest cent.

19. shoes: $70, 10% off

20. artwork: $250, 20% off

21. speakers: $180, 30% off

22. bicycle: $320, 25% off

23. antique chest: $179, 15% off

24. pendant: $93.50, 5% off

25. sofa: $749.95, 35% off

26. oven: $535.99, 20% off

27. guitar: $488.20, 25% off

28. weight machine: $919.70, 10% off

Lesson 5-8

5-9 Skills Practice

Simple Interest

Find the simple interest to the nearest cent.

1. $500 at 4% for 2 years

2. $800 at 9% for 4 years

3. $350 at 6.2% for 3 years

4. $280 at 5.5% for 4 years

5. $740 at 3.25% for 2 years

6. $1,150 at 7.6% for 5 years

7. $725 at 4.3% for $2\frac{1}{2}$ years

8. $266 at 5.2% for 3 years

9. $955 at 6.75% for $3\frac{1}{4}$ years

10. $1,245 at 5.4% for 4 years

11. $1,540 at 8.25% for 2 years

12. $2,180 at 7.7% for $2\frac{1}{2}$ years

13. $3,500 at 4.2% for $1\frac{3}{4}$ years

14. $2,650 at 3.65% for $4\frac{1}{2}$ years

Find the total amount in each account to the nearest cent.

15. $200 at 5% for 3 years

16. $700 at 6% for 2 years

17. $850 at 4% for 3 years

18. $350 at 8% for 2 years

19. $540 at 2.75% for 4 years

20. $360 at 4.5% for 5 years

21. $446 at 2.5% for 4 years

22. $780 at 3.6% for 3 years

23. $840 at 5.75% for $2\frac{1}{2}$ years

24. $530 at 7.25% for $1\frac{3}{4}$ years

25. $1,400 at 6.5% for 2 years

26. $1,880 at 4.3% for $3\frac{1}{2}$ years

27. $2,470 at 5.5% for 4 years

28. $3,200 at 9.75% for $1\frac{1}{2}$ years

29. $2,810 at 3.95% for $2\frac{1}{4}$ years

30. $4,340 at 8.12% for $3\frac{1}{4}$ years

6-1 Skills Practice

Line and Angle Relationships

Find the value of x in each figure.

1.
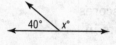
120° $x°$

2.
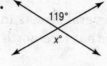
119°
$x°$

3.
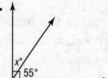
$x°$
55°

4.
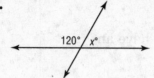
40° $x°$

5.
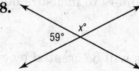
80° $x°$

6.
98° $x°$

7.
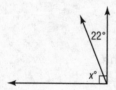
22°
$x°$

8.
59° $x°$

9.
$x°$ 6°

10.

89°
$x°$

11.
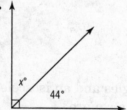
$x°$
44°

12.
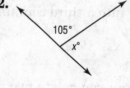
105°
$x°$

For Exercises 13 and 14, use the figure at the right.

13. Find the measure of angle 2. Explain your reasoning.

14. Find the measure of angle 6. Explain your reasoning.

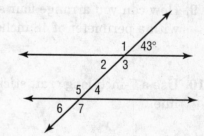

15. Angles Q and R and complementary.
    Find $m\angle R$ if $m\angle Q = 24°$.

16. Find $m\angle J$ if $m\angle K = 29°$ and
    $\angle J$ and $\angle K$ are supplementary.

Lesson 6-1

6-2 Skills Practice

Problem-Solving Investigation: Use Logical Reasoning

For Exercises 1–6, state whether the example uses *deductive* reasoning or *inductive* reasoning.

1. After checking the house numbers on several streets in your neighborhood, you discover that houses that face north always have an odd house number.

2. You determine the type of shape that a sticker is by examining its sides and angles.

3. You use a set of clues about how students received higher or lower scores on a math test as compared with other students to place the students in order from lowest grade to highest grade.

4. You roll a number cube 1,000 times and discover that it lands on the number 4 twice as many times as the number 1.

5. You find a way to use 2 larger containers to measure out the exact amount for a smaller container.

6. You determine what types of shapes will be created by connecting the corners of a regular hexagon.

For Exercises 7–10, solve each problem using logical reasoning.

7. Use a 5-liter container and a 3-liter container to measure out 4 liters of water into a third container.

8. How can you create two right triangles and an isosceles trapezoid by drawing two straight lines through a square?

9. How can you arrange four squares with 6-inch sides to create a figure with a perimeter of 48 inches?

10. Use a 7-inch-long craft stick and a 4-inch-long eraser to draw a 10-inch line.

Lesson 6-3

6-3 Skills Practice

Polygons and Angles

Find the sum of the measures of the interior angles of each polygon.

1. 13-gon

2. 17-gon

3. 18-gon

4. 24-gon

5. 32-gon

6. 35-gon

7. 21-gon

8. 29-gon

9. 54-gon

10. 64-gon

11. 81-gon

12. 150-gon

Find the measure of one interior angle of the given regular polygon. Round to the nearest hundredth if necessary.

13. heptagon (7-sided)

14. 26-gon

15. decagon (10-sided)

16. 23-gon

17. 37-gon

18. 51-gon

19. 48-gon

20. 85-gon

21. 72-gon

22. 49-gon

23. 66-gon

24. 500-gon

6-4 Skills Practice

Congruent Polygons

Determine whether the polygons shown are congruent. If so, name the corresponding parts and write a congruence statement.

1.

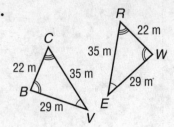

2.

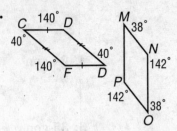

3.

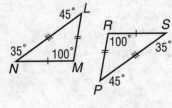

4.

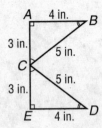

5.

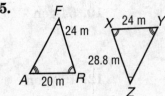

6.

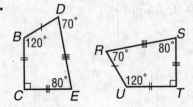

In the figure, △HFI ≅ △MLK. Find each measure.

7. m∠M

8. ML

9. m∠K

10. KM

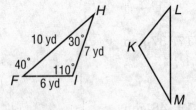

In the figure, quadrilateral ACDB ≅ quadrilateral EFGH. Find each measure.

11. m∠H

12. EF

13. m∠F

14. HG

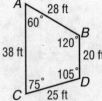
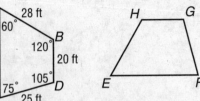

50

6-5 Skills Practice

Symmetry

For Exercises 1–12, complete parts a and b for each figure.

  a. Determine whether the figure has line symmetry. If it does, draw all lines of symmetry. If not, write *none*.

  b. Determine whether the figure has rotational symmetry. Write *yes* or *no*. If *yes*, name its angles of rotation.

1.

2.

3.

4.
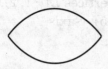

5.

6.

7.

8.

9.

10.

11.

12.
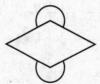

6-6 Skills Practice

Reflections

Draw the image of the figure after a reflection over the given line.

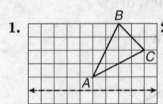

1.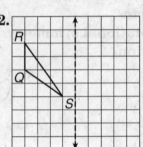

2.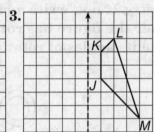

3.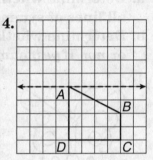

4.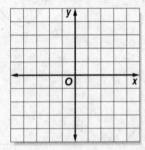

Graph the figure with the given vertices. Then graph the image of the figure after a reflection over the given axis and write the coordinates of its vertices.

5. triangle *ABC* with vertices *A*(1, 4), *B*(4, 1), and *C*(2, 5); *x*-axis

6. triangle *DEF* with vertices *D*(−1, 2), *E*(−3, 1), and *F*(−4, 5); *y*-axis

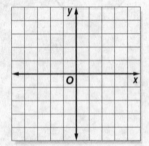

7. trapezoid *WXYZ* with vertices *W*(2, 4), *X*(2, −2), *Y*(4, −1), and *Z*(4, 3); *y*-axis

8. rhombus *QRST* with vertices *Q*(−1, 5), *R*(−4, 3), *S*(−1, 1), and *T*(2, 3); *x*-axis

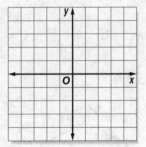

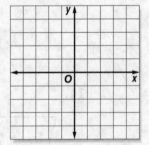

6-7 Skills Practice

Translations

Draw the image of the figure after the indicated translation.

1. 2 units left and 3 units up

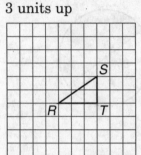

2. 4 units right and 1 unit up

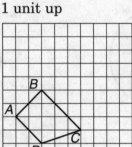

3. 1 unit left and 2 units down

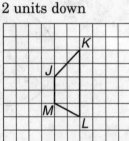

4. 5 units right and 3 units down

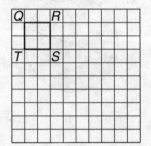

Graph the figure with the given vertices. Then graph the image of the figure after the indicated translation and write the coordinates of its vertices.

5. triangle *ABC* with vertices $A(-3, -1)$, $B(-4, -4)$, and $C(-1, -2)$ translated 4 units right and 1 unit up

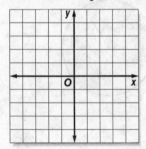

6. triangle *XYZ* with vertices $X(1, -2)$, $Y(3, -5)$, and $Z(4, 1)$ translated 5 units left and 3 units up

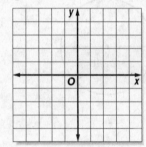

7. triangle *EFG* with vertices $E(1, 4)$, $F(-1, 1)$, and $G(2, -1)$ translated 3 units left and 1 unit down

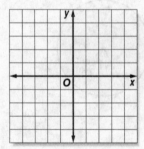

8. rhombus *WXYZ* with vertices $W(-4, 3)$, $X(-1, 1)$, $Y(2, 3)$, and $Z(-1, 5)$ translated 2 units right and 5 units down

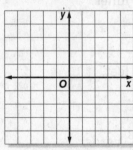

9. rectangle *QRST* with vertices $Q(-2, -4)$, $R(-2, 1)$, $S(-4, 1)$, and $T(-4, -4)$ translated 3 units right and 3 units up

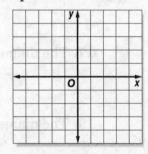

10. trapezoid *BCDE* with vertices $B(2, -1)$, $C(3, -3)$, $D(-3, -3)$, and $E(0, -1)$ translated 1 unit left and 4 units up

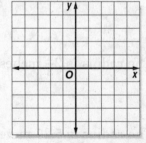

Lesson 6-7

7-1 Skills Practice

Circumference and Area of Circles

Find the circumference and area of each circle. Use 3.14 for π. Round to the nearest tenth.

1.
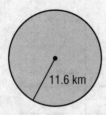
1 ft

2.
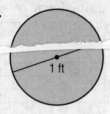
4 m

3.
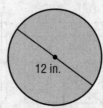
12 in.

4.
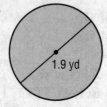
1.9 yd

5.
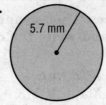
5.7 mm

6.
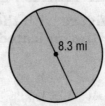
8.3 mi

7.
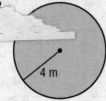
11.6 km

8.
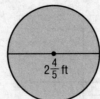
$2\frac{4}{5}$ ft

9.
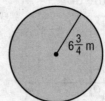
$6\frac{3}{4}$ m

10. The diameter is 7.7 feet.

11. The radius is 9.6 millimeters.

12. The radius is 3.8 meters.

13. The diameter is 17.4 yards.

14. The radius is 11.3 centimeters.

15. The diameter is $4\frac{3}{4}$ miles.

16. The radius is $2\frac{1}{3}$ inches.

17. The diameter is $7\frac{5}{8}$ feet.

18. The radius is 5.25 meters.

19. The diameter is $12\frac{3}{4}$ yards.

7-2 Skills Practice

Problem-Solving Investigation: Solve a Simpler Problem

For Exercises 1–3, rewrite the problem as a simpler problem.

1. Jerry has a square-shaped deep-dish pizza. What is the maximum number of pieces that can be made by using 6 cuts?

2. CDs come in packages of 25 and CD cases come in packages of 16. How many of each type of package will Lilly need to buy in order to make print 400 CDs and put them in cases with none left of either?

3. A restaurant has 10 triangular tables that can be pushed together in an alternating up-and-down pattern as shown below to form one long table for large parties. Each triangular table can seat 3 people per side. How many people can be seated at the combined tables?

For Exercises 4–15, rewrite to solve a simpler problem and solve. Find a reasonable answer.

4. 13×29

5. $48 + 32 + 87$

6. $74 \times (18 - 9)$

7. $33 \div 9$

8. $\dfrac{57}{113}$

9. $55 + 44 + 33$

10. 63×17

11. $532 - 389$

12. $78 \times 41 - 276$

13. $52 + 39 + 111$

14. $452 - 377$

15. $67 \times 34 \times 12$

Lesson 7-2

7-3 **Skills Practice**

Area of Composite Figures

Find the area of each figure. Use 3.14 for π. Round to the nearest tenth if necessary.

1.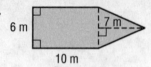
6 m
7 m
10 m

2.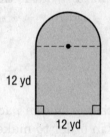
12 yd
12 yd

3.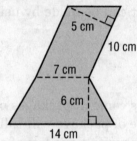
5 cm
10 cm
7 cm
6 cm
14 cm

4.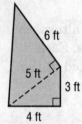
6 ft
5 ft
3 ft
4 ft

5.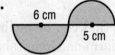
6 cm
5 cm

6.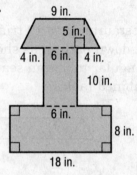
9 in.
5 in.
4 in. 6 in. 4 in.
10 in.
6 in.
18 in.
8 in.

7.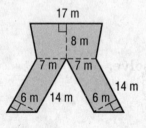
17 m
8 m
7 m 7 m
14 m
6 m 14 m 6 m

8.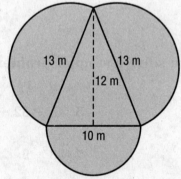
13 m 13 m
12 m
10 m

9.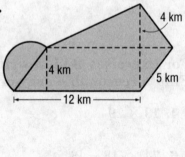
4 km
4 km
5 km
12 km

10. What is the area of a figure formed using a semicircle with a diameter of 16 feet and a trapezoid with a height of 8 feet and bases of 12 feet and 14 feet?

11. What is the area of a figure formed using a rectangle with a length of 13 kilometers and a width of 7 kilometers and a triangle with a base of 14 kilometers and a height of 11 kilometers?

7-4 Skills Practice

Three-Dimensional Figures

Identify each solid. Name the number and shapes of the faces. Then name the number of edges and vertices.

1.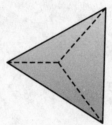

2.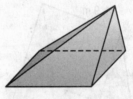

3.

4.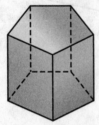

5.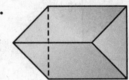

6.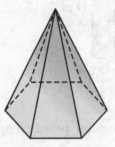

7.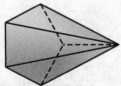

8.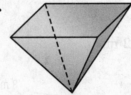

9.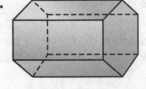

Lesson 7-4

7-5 Skills Practice

Volume of Prisms and Cylinders

Find the volume of each solid. Use 3.14 for π. Round to the nearest tenth if necessary.

1.

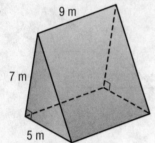

9 m
7 m
5 m

2.

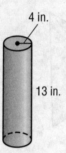

4 in.
13 in.

3.

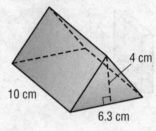

4 cm
10 cm
6.3 cm

4.

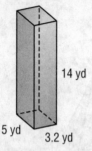

14 yd
5 yd
3.2 yd

5.

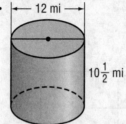

12 mi
$10\frac{1}{2}$ mi

6.

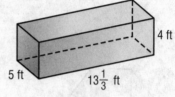

4 ft
5 ft
$13\frac{1}{3}$ ft

7. rectangular prism: length, 6 in.; width, 4 in.; height, 13 in.

8. triangular prism: base of triangle, 9 cm; altitude 1 cm; height of prism, 15 cm

9. rectangular prism: length, 3.6 mm; width, 4 mm; height, 15.5 mm

10. triangular prism: base of triangle, 6 yd; altitude 5.9 yd; height of prism, 12 yd

11. cylinder: diameter, 8 m; height, 16.2 m

12.

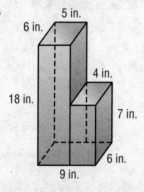

5 in.
6 in.
4 in.
18 in.
7 in.
6 in.
9 in.

13.

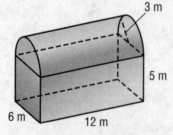

3 m
5 m
6 m
12 m

14.

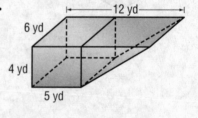

12 yd
6 yd
4 yd
5 yd

7-6 Skills Practice

Volume of Pyramids and Cones

Find the volume of each solid. Use 3.14 for π. Round to the nearest tenth if necessary.

1.

6 ft
2 ft

2.
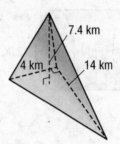
7.4 km
4 km
14 km

3.
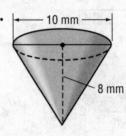
10 mm
8 mm

4.
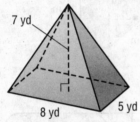
7 yd
8 yd
5 yd

5.
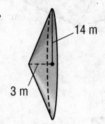
14 m
3 m

6.
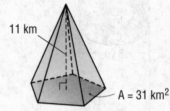
11 km
A = 31 km²

7.
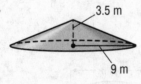
3.5 m
9 m

8.
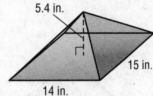
5.4 in.
15 in.
14 in.

9.
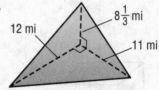
12 mi
$8\frac{1}{3}$ mi
11 mi

10. cone: diameter, 10 cm; height, 12 cm

11. triangular pyramid: triangle base, 20 mm; triangle height, 22 mm; pyramid height, 14 mm

12. triangular pyramid: triangle base, 19 in.; triangle height, 21 in.; pyramid height, 9 in.

13. cone: radius, 9.7 ft; height, 18 ft

Lesson 7-6

7-7 Skills Practice

Surface Area of Prisms and Cylinders

Find the lateral and total surface areas of each solid. Use 3.14 for π.
Round to the nearest tenth if necessary.

1.
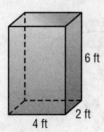
6 ft
2 ft
4 ft

2.
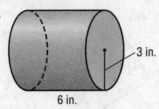
3 in.
6 in.

3.
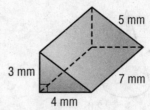
5 mm
3 mm
7 mm
4 mm

4.
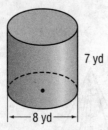
7 yd
8 yd

5.
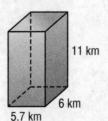
5 cm
6.1 cm
6.1 cm
7 cm
17 cm

6.
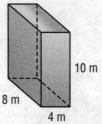
10 m
8 m
4 m

7.
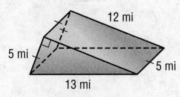
12 mi
5 mi
5 mi
13 mi

8.
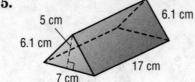
11 km
6 km
5.7 km

9.
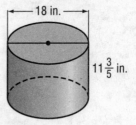
10.4 ft
9 ft

10.
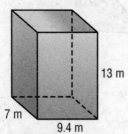
13 m
7 m
9.4 m

11.
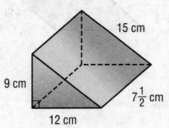
15 cm
9 cm
$7\frac{1}{2}$ cm
12 cm

12.
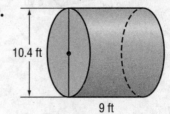
18 in.
$11\frac{3}{5}$ in.

13. cube: edge length, 11 m

14. rectangular prism: length, 9 cm; width, 13 cm; height, 18.4 cm

15. cylinder: radius, 9.4 mm; height, 15 mm

16. cylinder: diameter, 28 in.; height, 12.6 in.

7-8 Skills Practice

Surface Area of Pyramids

Find the surface area of each solid. Round to the nearest tenth if necessary.

1.

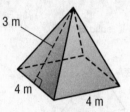

3 m

4 m

4 m

2.

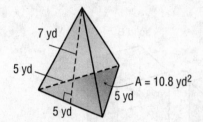

7 yd

5 yd

5 yd

5 yd

A = 10.8 yd²

3.

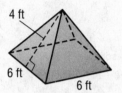

4 ft

6 ft

6 ft

4.

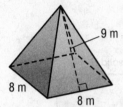

9 m

8 m

8 m

5.

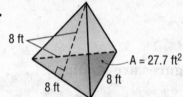

8 ft

8 ft

8 ft

A = 27.7 ft²

6.

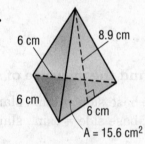

6 cm

8.9 cm

6 cm

6 cm

A = 15.6 cm²

7.

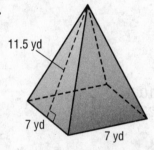

11.5 yd

7 yd

7 yd

8.

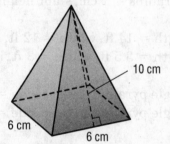

10 cm

6 cm

6 cm

9. square pyramid: base side length, 4 cm; slant height, 7.3 cm

10. square pyramid: base side length, 5 yd; slant height, 12.7 yd

Lesson 7-8

7-9 Skills Practice

Similar Solids

For Exercises 1–4, each pair of solids is similar. Find the volume of solid B.

1. solid A   solid B

    $V = 8$ units^3

    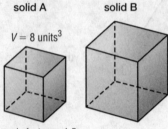

    scale factor = 1.5

2. solid A   solid B

    $V = 320$ units^3

    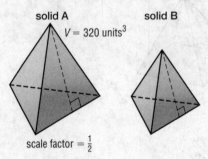

    scale factor = $\frac{1}{2}$

3. solid A   solid B

    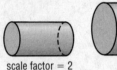

    scale factor = 2
    $V = 4\pi$ cubic units^3

4. solid A   solid B

    $V = 324\pi$ units^3

    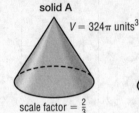

    scale factor = $\frac{2}{3}$

For Exercises 5–12, find the measure of x. All pairs of figures are similar.

5. square pyramid A: base side = 6 in., slant height = 21 in.
    square pyramid B: base side = x in., slant height = 7 in.

6. cone A: base radius = 8 cm, slant height = 20 cm
    cone B: base radius = x cm, slant height = 15 cm

7. prism A: length = 14 ft, width = 12 ft, height = 6 ft
    prism B: length = 3.5 ft, width = 3 ft, height = x ft

8. regular triangle pyramid A: base side = 3 in., slant height = 10 in.
    regular triangle pyramid B: base side = x in., slant height = 25 in.

9. cylinder A: base radius = 13 cm, length = 8 cm
    cylinder B: base radius = x cm, length = 24 cm

10. prism A: length = 7 ft, width = 15 ft, height = 8 ft
    prism B: length = 21 ft, width = x ft, height = 24 ft

11. square pyramid A: base side = 5 in., slant height = 18 in.
    square pyramid B: base side = x in., slant height = 9 in.

12. cone A: base radius = 16 m, height = 28 m
    cone B: base radius = x m, height = 21 m

8-1 Skills Practice

Simplifying Algebraic Expressions

Use the Distributive Property to rewrite each expression.

1. $4(j + 4)$

2. $5(n + 2)$

3. $(c + 9)3$

4. $2(w - 8)$

5. $(s - 7)7$

6. $-4(e + 6)$

7. $(b + 3)(-7)$

8. $-8(v - 7)$

9. $(2n + 3)6$

10. $5(c + d)$

11. $-7(3x - 1)$

12. $(e - f)3$

13. $2(-3m + 1)$

14. $(2b - 3)(-9)$

15. $-5(s + 7)$

16. $(t + 7)3$

17. $6(-2v + 4)$

18. $(m - n)(-3)$

Identify the terms, like terms, coefficients, and constants in each expression.

19. $4e + 7e + 5$

20. $5 - 4x - 8$

21. $-3h - 2h + 6h + 9$

22. $7 - 5y + 2 + 1$

23. $9k + 7 - k + 4$

24. $4z + 3 - 2z - z$

Simplify each expression.

25. $3t + 6t$

26. $4r + r$

27. $7f - 2f$

28. $9a - 8a$

29. $5c + 8c$

30. $2g - 5g$

31. $8k + 3 + 4k$

32. $7m - 5m - 6$

33. $9 - 6x + 5$

34. $7p - 1 - 9p + 5$

35. $-b - 3b + 8b + 4$

36. $5h - 6 - 8 + 7h$

37. $8b + 6 - 8b + 1$

38. $t - 5 - 2t + 5$

39. $4w - 5w + w$

40. $6m - 7 + 2m + 7$

41. $5f - 7f + f$

42. $12y - 8 + 4y + y$

43. $9a + 5 - 7a - 2a$

44. $6g - 7g + 13$

45. $7x + 6 - 9x - 3$

8-2   Skills Practice

Solving Two-Step Equations

Solve each equation. Check your solution.

1. $3n + 4 = 7$

2. $9 = 2s + 1$

3. $4c - 6 = 2$

4. $-4 = 2t - 2$

5. $3f - 12 = -3$

6. $8 = 4v + 12$

7. $5d - 6 = 9$

8. $2k + 12 = -4$

9. $-5 = 3m - 14$

10. $0 = 8z + 8$

11. $9a - 2 = -2$

12. $-8 + 4s = -16$

13. $-1 = 4 - 5x$

14. $5 = 9 - 2x$

15. $-2x + 12 = 14$

16. $1 - x = 8$

17. $-2 = -x + 4$

18. $11 = 2 - 3x$

19. $12 - 3x = 6$

20. $-6x + 5 = 17$

21. $13 = 18 - 5x$

22. $4x + 2x + 2 = 26$

23. $-18 = 9y - 5y + 10$

24. $-24 = 6a - 15 - 5a$

25. $3z - 17 + 2z = 13$

26. $22 = 4 + 8e - 2e$

27. $-15 = 9r + 1 - 7r$

28. $8k - 8 + k = 10$

29. $-27 = 2c - 7 - 6c$

30. $11 = 18 + 3f + 4f$

8-3　Skills Practice

Writing Two-Step Equations

Translate each sentence into an equation. Then find each number.

1. Four more than twice a number is 8.

2. Three more than four times a number is 15.

3. Five less than twice a number is 7.

4. One less than four times a number is 11.

5. Seven more than the quotient of a number and 2 is 10.

6. Six less than six times a number is 12.

7. Five less than the quotient of a number and 3 is −7.

8. Seven more than twice a number is 1.

9. The difference between 5 times a number and 3 is 12.

10. Nine more than three times a number is −6.

11. Nine more than the quotient of a number and 4 is 12.

12. Four less than the quotient of a number and 3 is −10.

13. Nine less than six times a number is −15.

14. Three less than the quotient of a number and 6 is 1.

15. Eight more than the quotient of a number and 5 is 3.

16. The difference between twice a number and 11 is −23.

8-4 Skills Practice

Solving Equations with Variables on Each Side

Solve each equation. Check your solution.

1. $3w + 6 = 4w$

2. $a + 18 = 7a$

3. $8c = 5c + 21$

4. $11d + 10 = 6d$

5. $2e = 4e - 16$

6. $7v = 2v - 20$

7. $4n - 6 = 10n$

8. $2y + 27 = 5y$

9. $8h = 6h - 14$

10. $18 - 2g = 4g$

11. $4x - 9 = 6x - 13$

12. $5c - 15 = 2c + 6$

13. $t + 10 = 7t - 14$

14. $8z + 6 = 7z + 4$

15. $2e - 12 = 7e + 8$

16. $9k + 6 = 8k + 13$

17. $2d + 10 = 6d - 10$

18. $-2a - 9 = 6a + 15$

19. $8 - 3k = 3k + 2$

20. $7t - 4 = 10t + 14$

21. $3c - 15 = 17 - c$

22. $14 + 3n = 5n - 6$

23. $3y + 5.2 = 2 - 5y$

24. $10b - 2 = 7b - 7.4$

25. $2m - 2 = 6m - 4$

26. $3g + 5 = 7g + 4$

27. $4s - 1 = 8 - 2s$

28. $9w + 3 = 4w - 9$

29. $6z - 7 = 2z - 2$

30. $3 - a = 4a + 12$

8-5 Skills Practice

Problem-Solving Investigation: Guess and Check

Use the guess and check strategy to solve each problem.

1. **NUMBER THEORY** A number cubed is 1,728. What is the number?

2. **MONEY** Jackson has exactly $43 in $1, $5, and $10 bills. If he has 8 bills, how many of each bill does he have?

3. **NUMBERS** Jona is thinking of two numbers. One number is 18 more than twice the other number. The sum of the numbers is 48. What two numbers is Jona thinking of?

4. **PACKAGES** The packages in a mail driver's truck weigh a total of 950 pounds. The large packages weigh 20 pounds each and the small packages weigh 10 pounds each. If he has 10 more large packages than small packages, how many large and small packages are on the truck?

5. **NUMBER THEORY** One number is twice the other. The sum of the numbers is 246. What are the two numbers?

6. **MOVIE RENTALS** A movie rental store rented 3 times as many DVDs as videos. DVDs rent for $5 a day and videos rent for $3 a day. If the total rental income for one day was $2,160, how many DVDs and videos did the store rent?

Lesson 8-5

8-6 Skills Practice

Inequalities

Write an inequality for each sentence.

1. **SPORTS** You need to score at least 30 points to take the lead.

2. **SEASONS** There are less than 12 hours of daylight each day in winter.

3. **TRAVEL** The bus seats at most 60 people.

4. **MONEY** The coupon is good for any item that costs less than $10.

5. **TESTS** A score of at least 92 on the test is considered an A.

6. **HEALTH** The baby weighed more than 7 pounds at birth.

7. **DRIVING** Victor drives less than 12,000 miles per year.

8. **TRAVEL** Your waiting time will be 18 minutes or less.

9. **SCHOOL TRIPS** At least 15 students must sign up for the school trip.

For the given value, state whether each inequality is *true* or *false*.

10. $y + 2 < 8, y = 3$

11. $12 > u - 1, u = 14$

12. $p + 5 \geq -6, p = 1$

13. $-6 < a - 3, a = -1$

14. $4s \leq 15, s = 4$

15. $-5 > 1 - d, d = -9$

16. $-2 - g \geq -7, g = 5$

17. $\dfrac{k}{3} > 4, k = 12$

18. $4 < \dfrac{-10}{z}, z = -2$

19. $\dfrac{12}{m} \geq 3, m = 4$

Graph each inequality on a number line.

20. $v \geq 3$

21. $b > 5$

22. $n \leq -3$

23. $w < 4$

24. $r > -1$

25. $h \geq -7$

68

8-7 Skills Practice

Solving Inequalities by Adding or Subtracting

Solve each inequality. Check your solution.

1. $r + 5 < 6$ 2. $e - 3 > 2$ 3. $-8 \geq k - 5$

4. $y + 6 > 5$ 5. $n - 4 \geq 6$ 6. $-4 > g - 10$

7. $-1 \leq m + 8$ 8. $t + 1 \leq 6$ 9. $-17 > u - 2$

10. $5 + x \leq -7$ 11. $10 > p + 9$ 12. $-4 + z < -12$

13. $5 \leq q + 8$ 14. $k - 6 > 2$ 15. $s + 7 \leq -13$

Write an inequality and solve each problem.

16. Two more than a number is less than eleven.

17. Five less than a number is at least -2.

18. The difference between a number and 6 is no more than 5.

19. The sum of a number and 7 is more than 1.

20. The difference between a number and ten is greater than 9.

21. Four less than a number is less than 11.

Solve each inequality and check your solution. Then graph the solution on a number line.

22. $9 < p - 6$

23. $w + 4 \geq -3$

24. $1 > z + 5$

25. $-6 \leq s - 7$

26. $b - 3 \leq 7$

27. $v + 9 > 23$

28. $4 + v \geq 5$

29. $m + 7 < 11$

Lesson 8-7

8-8 Skills Practice

Solving Inequalities by Multiplying or Dividing

Solve each inequality and check your solution. Then graph the solution on a number line.

1. $2v > 10$

+—+—+—+—+—+—+—+—+—+—+—+

2. $\frac{p}{3} < -21$

+—+—+—+—+—+—+—+—+—+—+—+

3. $-12 \le 4g$

-6 -5 -4 -3 -2 -1 0 1 2

4. $60 \ge 5c$

6 7 8 9 10 11 12 13 14

5. $\frac{a}{2} > -2$

-6 -5 -4 -3 -2 -1 0 1 2

6. $1 \le \frac{u}{6}$

0 1 2 3 4 5 6 7 8

7. $-14 > 14n$

+—+—+—+—+—+—+—+—+—+—+—+

8. $-4d \ge -28$

+—+—+—+—+—+—+—+—+—+—+—+

Solve each inequality. Check your solution.

9. $3a + 2 < -4$

10. $5b - 4 \ge -29$

11. $\frac{m}{4} + 6 < 10$

12. $-7d + 8 \le 1$

13. $\frac{z}{-8} - 5 < -2$

14. $2 + \frac{r}{6} > -1$

15. $4v - 6 \le 2$

16. $3 + \frac{h}{-7} \ge 1$

17. $-2y - 5 \le 19$

Write an inequality for each sentence. Then solve the inequality.

18. Six times a number is less than 60.

19. The quotient of a number and 2 is more than -11.

20. The quotient of a number and 5 is at most 25.

21. Two times a number is more than 36.

22. Negative three times a number is at least -60.

23. Four times a number is greater than -56.

9-3 Skills Practice

Representing Linear Functions

Complete the function table. Then graph the function.

1. $y = x + 4$

x	$x + 4$	y	(x, y)
-2			
-1			
0			
1			

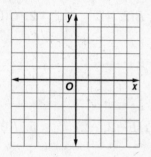

2. $y = 2x - 1$

x	$2x - 1$	y	(x, y)
-1			
0			
1			
2			

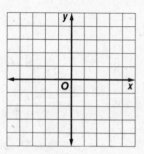

Graph each function.

3. $y = x - 6$

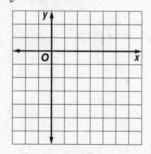

4. $y = 2x - 3$

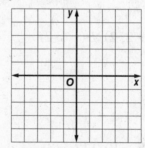

5. $y = 1 - x$

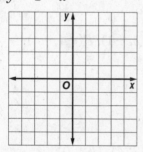

6. $y = 3x + 2$

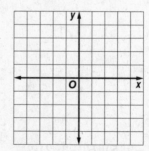

7. $y = \dfrac{x}{2} + 2$

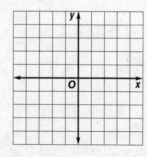

8. $y = \dfrac{x}{3} - 1$

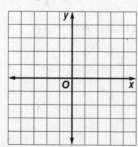

Lesson 9-3

9-4 Skills Practice

Slope

Find the slope of the line that passes through each pair of points.

1. $A(-2, -4)$, $B(2, 4)$ **2.** $C(0, 2)$, $D(-2, 0)$ **3.** $E(3, 4)$, $F(4, -2)$

4. $G(-3, -1)$, $H(-2, -2)$ **5.** $I(0, 6)$, $J(-1, 1)$ **6.** $K(0, -2)$, $L(2, 4)$

7. $O(1, -3)$, $P(2, 5)$ **8.** $Q(1, 0)$, $R(3, 0)$ **9.** $S(0, 4)$, $T(1, 0)$

10. $U(1, 3)$, $V(1, 5)$ **11.** $W(2, -2)$, $X(-1, 1)$ **12.** $Y(-5, 0)$, $Z(-2, -4)$

13. $A(2, -1)$, $B(-4, -4)$ **14.** $C(-2, 2)$, $D(-4, 2)$ **15.** $E(-1, -4)$, $F(-3, 0)$

16. $G(7, 4)$, $H(2, 0)$ **17.** $K(2, -2)$, $L(2, -3)$ **18.** $M(-1, -1)$, $N(-4, -5)$

19. $O(5, -3)$, $P(-3, 4)$ **20.** $Q(-1, -3)$, $R(1, 2)$ **21.** $W(3, -5)$, $X(1, 1)$

22. $Y(2, 2)$, $Z(-5, -4)$ **23.** $C(0, -2)$, $D(3, -2)$ **24.** $G(-3, 5)$, $H(-3, 2)$

9-5 Skills Practice

Direct Variation

For Exercises 1–3, determine whether each linear function is a direct variation. If so, state the constant of variation.

1.

Price x	$5	$10	$15	$20
Tax y	$0.41	$0.82	$1.23	$1.64

2.

Hours x	11	12	13	14
Distance y (miles)	154	167	180	194

3.

Age x	8	9	10	11
Grade y	3	4	5	6

For Exercises 4–12, y varies directly with x. Write an equation for the direct variation. Then find each value.

4. If $y = 8$ when $x = 3$, find y when $x = 45$.

5. If $y = -4$ when $x = 10$, find y when $x = 2$.

6. If $y = 27$ when $x = 8$, find y when $x = 11$.

7. Find y when $x = 12$ if $y = 2$ when $x = 5$.

8. Find y when $x = 3$ if $y = -4$ when $x = -9$.

9. Find y when $x = -6$ if $y = 15$ when $x = -5$.

10. If $y = 20$ when $x = 8$, what is the value of x when $y = -2$?

11. If $y = -30$ when $x = 15$, what is the value of x when $y = 60$?

12. If $y = 42$ when $x = 15$, what is the value of x when $y = 70$?

Lesson 9-5

9-6 Skills Practice

Slope-Intercept Form

State the slope and *y*-intercept of the graph of each equation.

1. $y = x + 4$

2. $y = 2x - 2$

3. $y = 3x - 1$

4. $y = -x + 3$

5. $y = \frac{1}{2}x - 5$

6. $y = -\frac{1}{3}x + 4$

7. $y - 2x = -1$

8. $y + 4x = 2$

9. $y = \frac{3}{2}x - 3$

10. Graph a line with a slope of 1 and a *y*-intercept of −4.

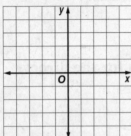

11. Graph a line with a slope of 2 and a *y*-intercept of −3.

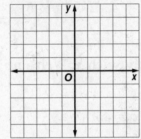

12. Graph a line with a slope of $\frac{1}{3}$ and a *y*-intercept of 1.

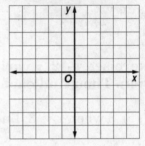

Graph each equation using the slope and *y*-intercept.

13. $y = 3x - 3$

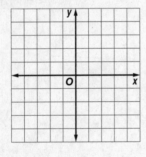

14. $y = -x + 1$

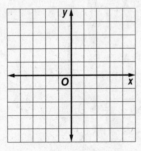

15. $y = \frac{1}{2}x - 2$

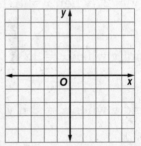

16. $y = 4x - 2$

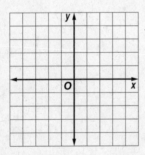

17. $y = -\frac{3}{2}x + 1$

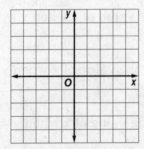

18. $y = \frac{2}{3}x - 3$

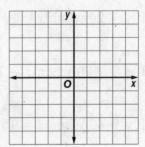

9-9 Skills Practice

Scatter Plots

Explain whether the scatter plot of the data for the following shows a positive, negative, or no relationship.

1. rotations of a bicycle tire and distance traveled on the bicycle

2. number of pages printed by an inkjet printer and the amount of ink in the cartridge

3. age of a child and the child's shoe size

4. number of letters in a person's first name and the person's height

5. shots attempted and points made in a basketball game

6. year and winning time in the 100-meter dash in the Olympics

7. diameter of the trunk of a tree and the height of the tree

8. number of a bank account and the amount of money in the bank account

9. length of a taxi ride and the amount of the fare

10. daily high temperature and the amount of clothing a person wears

11. a person's age and a person's street address

12. outside temperature and the cost of air conditioning

13. the age of a car and how many people fit inside of it

14. inches of rainfall in the last 30 days and the water level in a reservoir

15. miles ridden on a bicycle tire and thickness of the tire tread

16. population of a U.S. state and the number of U.S. senators a state has

10-1 Skills Practice

Linear and Nonlinear Functions

Determine whether each graph, equation, or table represents a *linear* or *nonlinear* function. Explain.

1.
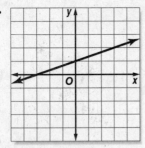

2.
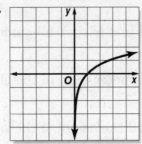

3.
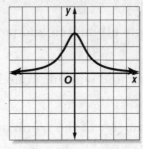

4.
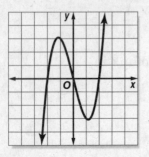

5.
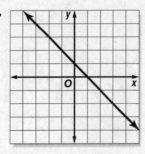

6.
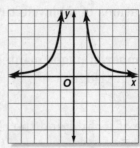

7. $y = 2x$

8. $y = 3x^2 + 5$

9. $y = \dfrac{6}{x}$

10. $y = x^3 + 7$

11. $y = -6$

12. $y = -\dfrac{5x}{2}$

13.

x	1	2	3	4
y	5	7	9	11

14.

x	−2	0	2	4
y	0	1	3	9

15.

x	−1	0	1	2
y	8	4	0	−4

16.

x	2	3	4	5
y	3	5	8	12

17.

x	−2	1	4	7
y	−4	1	6	11

18.

x	3	6	9	12
y	10	6	3	1

10-2 **Skills Practice**

Graphing Quadratic Functions

Graph each function.

1. $y = -4x^2$

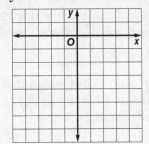

2. $y = 1.5x^2$

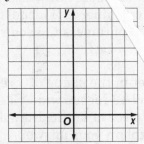

3. $y = x^2 + 4$

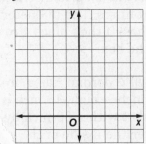

4. $y = x^2 - 5$

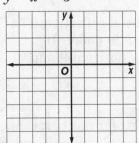

5. $y = -x^2 + 3$

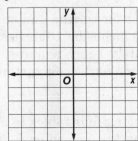

6. $y = -x^2 - 1$

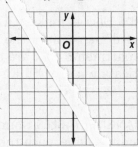

7. $y = 2x^2 - 3$

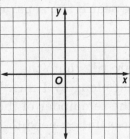

8. $y = -2x^2 + 1$

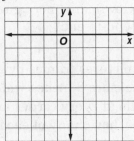

9. $y = -2x^2 - 2$

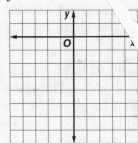

10. $y = 3x^2 + 1$

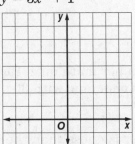

11. $y = -3x^2 + 3$

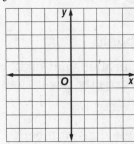

12. $y = 0.5x^2 + 2$

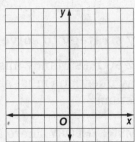

13. $y = 1.5x^2 - 1$

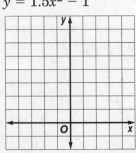

14. $y = 2.5x^2 + 1$

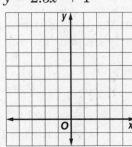

15. $y = -0.5x^2 - 1$

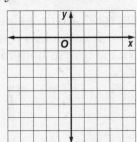

Lesson 10-2

10-3 Skills Practice

Problem-Solving Investigation: Make a Model

Make a model to solve each problem.

1. **SHIPPING** A spice distributor is making boxes in which to pack cylindrical spice containers. The diameter of each container is 2 inches. The height of each container is 4 inches. If they place 4 rows with 3 containers in each row in a box, what is the volume of the box?

2. **SEWING** Jordan has a bread basket in the shape of a rectangular prism that measures 12 inches high, 18 inches long, and 16 inches wide. She wants to cover the inside of the basket with a 50-inch by 20-inch piece of fabric. Does Jordan have enough fabric to cover the inside of the basket? Explain your answer.

3. **BEADS** Elsa is making a wooden box for sorting and storing her bead collection. The outer dimensions of the box are 10 inches by 10 inches. She wants to make 100 compartments that are approximately 1-inch squares. How many horizontal and vertical dividers will Elsa need to make the compartments?

4. **ARRANGING TABLES** Donna is arranging four tables to make seating for her party guests. Standing alone, each table will seat 4 people on each side and 2 people at each end. She can either place the tables end-to-end to make one long table or she can separate the tables into four individual tables. How many more guests can she seat if she separates the tables than if she places them end-to-end?

5. **MAKING FRAMES** Julian is making pictures frames by gluing square tiles onto the wooden sides. The wooden sides measure 8 inches wide by 10 inches long by 1 inch deep. If he glues a 1-inch square tile at every corner and covers the remainder of the wood sides with $\frac{1}{2}$-inch square tiles, how many of each size tile does Julian need to make 4 frames?

Use any strategy to solve each problem.

6. **QUIZ SCORES** Mandy answered 10 questions out of 12 correctly on her math quiz. How many questions must she answer correctly to get the same score on a quiz with 30 questions?

7. **NUMBER THEORY** There are two single digit numbers. One number is 4 less than the other number. The sum of the digits is 12. Find the two numbers.

8. **GARDENING** Justin helped his dad in the yard 3 times as long as Paula. Paula helped her dad 2 hours less than Carly. Carly helped her dad in the yard 4 hours. How many hours did Justin help his dad?

10-4 Skills Practice

Graphing Cubic Functions

Graph each function.

1. $y = 2x^3 + 1$

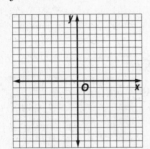

2. $y = -2x^3$

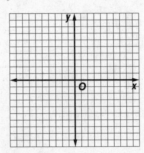

3. $y = x^3 - 3$

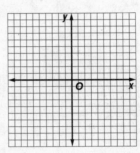

4. $y = -3x^3$

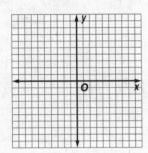

5. $y = -x^3 - 2$

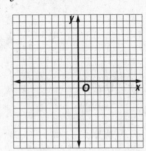

6. $y = 2x^3 - 2$

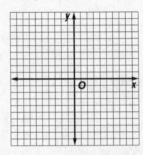

7. $y = x^3 + 3$

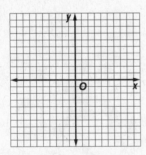

8. $y = -3x^3 - 2$

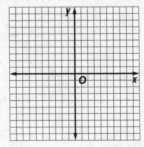

9. $y = -x^3 + 1$

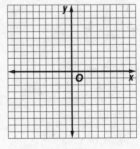

10. $y = -2x^3 + 2$

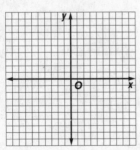

11. $y = -2x^3 - 2$

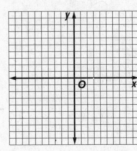

12. $y = x^3 + 4$

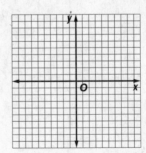

Lesson 10-4

10-5 Skills Practice

Multiplying Monomials

Multiply. Express using exponents.

1. $2^7 \cdot 2^2$

2. $4^2 \cdot 4^4$

3. $10^2 \cdot 10^3$

4. $k^8 \cdot k$

5. $t^7 \cdot t^6$

6. $2w^2 \cdot 5w^2$

7. $3e^3 \cdot 7e^3$

8. $4r^4(-4r^3)$

9. $(-6t^7)(5t^2)$

10. $7y^3 \cdot 6y$

11. $(3u^5)(-9u^6)$

12. $(-2p^7)(-8p^3)$

13. $(5c^4)(-7c)$

14. $(8z^7)(3z^6)$

15. $(-3l^2w^3)(2lw^4)$

16. $10c^2 \cdot c^2d^6$

17. $(-11w^4)(-5w^3x^4)$

18. $q^2r^3(3q)$

19. $(8f^6)(-6f^2g^5)$

20. $(10d^2)(-5d^5)$

21. $9k^2(-k^2l^5)$

22. $(-4b^6)(-b^2c^3)$

23. $(10t^4v^5)(3t^2v^5)$

24. $a^4c^6(a^2c)$

10-6 Skills Practice

Dividing Monomials

Lesson 10-6

Multiply or divide. Express using exponents.

1. $\dfrac{2^9}{2^3}$

2. $\dfrac{3^8}{3^4}$

3. $\dfrac{5^9}{5^2}$

4. $\dfrac{8^7}{8}$

5. $\dfrac{b^{12}}{b^5}$

6. $\dfrac{12n^5}{4n^2}$

7. $\dfrac{14m^3}{7m^2}$

8. $\dfrac{9r^8}{3r^4}$

9. $\dfrac{24t^9}{6t^3}$

10. $\dfrac{18y^6}{2y}$

11. $\dfrac{a^4c^6}{a^2c}$

12. $\dfrac{15x^8y^4}{3x^5y^2}$

13. $\dfrac{-21s^6t^3}{3s^2t}$

14. $\dfrac{34v^7}{2v^3}$

15. $\dfrac{4^2q^5}{2q^2}$

16. $\dfrac{5^{10}}{5^2}$

17. $\dfrac{7^9}{7^6}$

18. $\dfrac{r^8}{r^7}$

19. $\dfrac{(-y)^7}{(-y)^2}$

20. $\dfrac{g^{-12}}{g^8}$

21. $\dfrac{8^2}{8^{-4}}$

22. $\dfrac{7^9}{7^6}$

23. $\dfrac{24x^7}{6x^2}$

24. $\dfrac{15t^{-2}}{3t}$

10-7 Skills Practice

Powers of Monomials

Simplify.

1. $(7^2)^3$

2. $(3^2)^6$

3. $(8^3)^2$

4. $(9^4)^2$

5. $(d^7)^6$

6. $(m^5)^5$

7. $(h^6)^3$

8. $(z^7)^3$

9. $[(4^3)^2]^2$

10. $(-5a^2b^7)^7$

11. $(2m^5g^{11})^6$

12. $[(2^3)^3]^2$

13. $(7a^5b^6)^4$

14. $(7m^3n^{11})^5$

15. $(-3w^3z^8)^5$

16. $(-7r^4s^{10})^4$

Express the area of each square below as a monomial.

17.

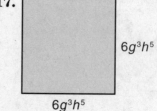

$6g^3h^5$

$6g^3h^5$

18.

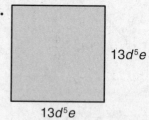

$13d^5e$

$13d^5e$

19.

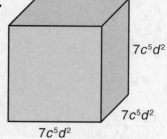

$7c^5d^2$

$7c^5d^2$

$7c^5d^2$

20.

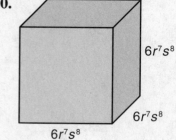

$6r^7s^8$

$6r^7s^8$

$6r^7s^8$

10-8 Skills Practice

Roots of Monomials

Lesson 10-8

Simplify.

1. $\sqrt{m^2}$

2. $\sqrt{x^6}$

3. $\sqrt{p^2r^4}$

4. $\sqrt{a^6b^8}$

5. $\sqrt{16n^4}$

6. $\sqrt{36w^{10}}$

7. $\sqrt{121x^8y^4}$

8. $\sqrt{225a^2b^8}$

9. $\sqrt{400m^6n^{14}}$

10. $\sqrt[3]{c^3}$

11. $\sqrt[3]{t^9}$

12. $\sqrt[3]{f^6g^{15}}$

13. $\sqrt[3]{v^{12}w^{18}}$

14. $\sqrt[3]{27g^{15}}$

15. $\sqrt[3]{8p^{24}}$

16. $\sqrt[3]{64k^{12}m^{18}}$

17. $\sqrt[3]{125x^3y^{12}}$

18. $\sqrt[3]{8a^{12}b^6c^{21}}$

Write a radical expression for each square root.

19. $4x^4y^2$

20. $8|a^3|b^4$

21. $12p^6|q^7|$

Write a radical expression for each cube root.

22. $5m^3n$

23. $7d^6g^9$

24. $2j^7k^5$

11-1 Skills Practice

Problem-Solving Investigation: Make a Table

Make a table to solve each problem.

1. **SCIENCE** Ecology students investigated the number of chirps a cricket makes in 15 seconds. Their results are shown below. What is the most common number of chirps made by crickets in a 15-second interval?

   30 31 30 32 32 31 30 30 30 31 30 32 31 30 30 31 32
   31 30 31 30 30 32 30 30 31 31 32 30 30 32 32 30 30

2. **SPORTS TRAINING** Thirty athletes were surveyed to determine how many hours per week they spend training for a marathon. Organize the data in a table using intervals 1–5, 6–10, 11–15, 16 or more. What is the most common interval of hours practiced in a week?

Interval	Tally	Frequency

   4 12 15 6 14 13 9 18 14 8

   13 4 11 13 11 2 17 7 14 15

   8 11 15 1 12 16 9 18 10 19

3. **BOOKS** Mr. Whitney's class listed the number of books each student read during the first grading period. The results are shown at the right. Find the number of books read that is listed most frequently.

0 3 6	5 6 3
2 8 4	3 3 4
7 5 3	7 8 2
2 9 6	9 7 4
7 5	1 0

4. **GAS PRICES** A local news station researched the price of gas at 20 gas stations throughout the state and recorded the following results. Organize the data in a table using intervals $1.99 or less, $2–$2.15, more than $2.15. What is the most common interval of gas prices?

   $2.05 $2.19 $2.18 $2.15 $2.19 $2.20 $2.29 $2.05 $1.99 $2.18
   $2.19 $2.08 $2.00 $2.16 $2.19 $1.99 $2.21 $2.20 $2.00 $2.16

Interval	Tally	Frequency

5. **ATTENDANCE** The number of days students in Ms. Roe's class were absent are as follows.

   1 0 3 4 1 0 2 0 3 4 1 3 4 1 2 0 1 2 0 3
   4 1 3 4 1 2 0 1 2 4 3 1 2 2 2 1 3 1 1 2

   What is the most frequent number of days absent?

11-2 Skills Practice

Histograms

1. BASKETBALL The frequency table at the right shows the average points per game for all NBA teams for a recent season. Draw a histogram to represent the set of data.

Average Points per Game for NBA Teams, a Recent Regular Season		
Points	**Tally**	**Frequency**
87–90.9	I	1
91–94.9	⊞ IIII	9
95–98.9	⊞ ⊞ I	11
99–102.9	⊞ II	7
103–106.9	I	1
107–110.9	I	1

2. GOLF The frequency table at the right shows the score of the winner of the Masters golf tournament for the years 1970–2006. Draw a histogram to represent the set of data.

Score of the Winner of the Masters from 1970–2006		
Score	**Tally**	**Frequency**
266–270	I	1
271–275	⊞	5
276–280	⊞ ⊞ ⊞ ⊞ II	22
281–285	⊞ III	8
286–290	I	1

3. RAINFALL The frequency table at the right shows the average annual precipitation for the 50 states. Draw a histogram to represent the set of data.

Average Annual Precipitation for the 50 States

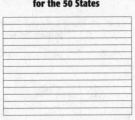

Average Annual Precipitation for the 50 States		
Precipitation (in.)	**Tally**	**Frequency**
0–11.9	IIII	4
12–23.9	⊞ IIII	9
24–35.9	⊞ III	8
36–47.9	⊞ ⊞ ⊞ ⊞ II	22
48–59.9	IIII	4
60–71.9	III	3

Lesson 11-2

11-3 Skills Practice

Circle Graphs

Construct a circle graph for each set of data.

1.

U.S. Energy Consumption	
Type	**Percent**
Commercial	18%
Transportation	28%
Residential	21%
Industrial	33%

U.S. Energy Consumption

2.

Type of Trucks Sold in U.S.	
Type	**Percent**
Compact Pickup	9%
Van	15%
Full-Size Pickup	27%
SUV	45%
Medium/Heavy	4%

Type of Trucks Sold in U.S.

3.

Davis Cup Winner, 1981–2006	
Country	**Wins**
Australia	4
France	3
Germany	3
Spain	2
Sweden	6
U.S.	5
Russia	2
Croatia	1

Davis Cup Winner, 1981–2006

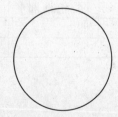

4.

Top 5 Largest American Indian Tribes	
Tribe	**Number (thousands)**
Cherokee	730
Navajo	298
Latin American Indian	181
Choctaw	159
Sioux	153

Top 5 Largest American Indian Tribes

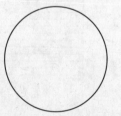

11-8 Skills Practice

Select an Appropriate Display

Select an appropriate type of display for each situation. Justify your reasoning.

1. energy usage in your home for the past year, categorized by month

2. exam scores for a whole class, arranged in intervals

3. sales of a leading brand of toothpaste for the last 10 years

4. average weight of a pet dog, categorized by breed

5. runs scored by individual members of a baseball team, as compared to the team total

6. ages of all 40 employees of a small company

Select an appropriate type of display for each situation. Justify your reasoning. Then construct a display.

7.

Points per Game by Shaquille O'Neal 1999–2007	
Season	**Points per Game**
99–00	29.7
00–01	28.7
01–02	27.2
02–03	27.5
03–04	21.5
04–05	22.9
05–06	23.4
06–07	19.6

8.

Share of Workers by Commute Time	
Commute Time	**Percent of Workers**
Less than 15 min	30%
15–29 min	36%
30–39 min	16%
40–59 min	11%
60 min or more	7%

Lesson 11-8

12-1 Skills Practice

Counting Outcomes

Draw a tree diagram to determine the number of outcomes.

1. A hat comes in black, red, or white, and medium or large.

2. You have a choice of peach or vanilla yogurt topped with peanuts, granola, walnuts, or almonds.

Use the Fundamental Counting Principle to find the number of possible outcomes.

3. A test consists of 5 true-false questions.

4. A number cube is rolled, and a dime and penny are tossed.

5. Canned beans are packed in 3 sizes and 7 varieties.

6. There are 5 choices for each of 6 multiple-choice questions on a history quiz.

12-2 **Skills Practice**

Probability of Compound Events

For Exercises 1–6, a number cube is rolled and the spinner at the right is spun. Find each probability.

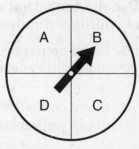

1. P(1 and A) **2.** P(odd and B)

3. P(prime and D) **4.** P(greater than 4 and C)

5. P(less than 3 and consonant) **6.** P(prime and consonant)

7. What is the probability of spinning the spinner above 3 times and getting a vowel each time?

8. What is the probability of rolling a number cube 3 times and getting a number less than 3 each time?

Each spinner at the right is spun. Find each probability.

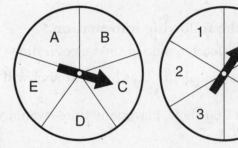

9. P(A and 2)

10. P(vowel and even)

11. P(consonant and 1)

12. P(D and greater than 1)

There are 3 red, blue, and 2 yellow marbles in a bag. Once a marble is selected, it is not replaced. Find each probability.

13. P(red and then yellow) **14.** P(blue and then yellow)

15. P(red and then blue) **16.** P(two yellow marbles)

17. P(two red marbles in a row) **18.** P(three red marbles)

GAMES There are 13 yellow cards, 6 blue, 10 red, and 8 green cards in a stack of cards turned face down. Once a card is selected, it is not replaced. Find each probability.

19. P(2 blue cards) **20.** P(2 red cards)

21. P(a yellow card and then a green card) **22.** P(a blue card and then a red card)

23. P(two cards that are not red) **24.** P(two cards that are neither red or green)

12-3 Skills Practice

Experimental and Theoretical Probability

Use the table that shows the results of rolling a number cube 50 times.

Result	Number of Times
1	6
2	10
3	8
4	7
5	10
6	9

1. Based on the results, what is the probability of getting a five?

2. Based on the results, how many fives would you expect to occur in 300 rolls?

3. What is the theoretical probability of getting a five?

4. Based on the theoretical probability, how many fives would you expect to occur in 300 rolls?

5. Compare the theoretical probability to the experimental probability.

ARCHERY Use the following information.

In archery class, Jocelyn missed the target 5 times in 40 shots.

6. What is the probability that her next shot will miss the target?

7. In her next 160 shots, how many times would you expect Jocelyn to miss the target?

PETS For Exercises 8–11, use the results of a survey of 200 people shown at the right.

First Pet	Number
bird	32
cat	56
dog	66
rabbit	19
other	27

8. What is the probability that a person says his or her first pet was a cat?

9. Out of 500 people, how many would you expect to say a cat was his or her first pet?

10. What is the probability that a person says his or her first pet was a bird?

11. Out of 500 people, how many would you expect to say a bird was their first pet?

12. **FIGURE SKATING** At figure skating practice, Michelle successfully landed 15 out of 18 attempts at a double axel. What is the experimental probability that she will successfully land a double axel?

12-4 Skills Practice

Problem-Solving Investigation: Act It Out

For Exercises 1–7, use the act it out strategy to solve.

1. A piece on a game board moves forward 8 spaces on its first turn and moves backward 3 spaces on its second turn. If the pattern continues, how many turns will it take for the piece to move at least 30 spaces?

2. How many ways can you arrange 3 paintings in a row on a wall?

3. How many different combinations of nickels, dimes, and pennies can be used to make $0.10?

4. A piece on a game board moves forward 6 spaces on its first turn and moves backward 5 spaces on its second turn. If the pattern continues, how many turns will it take for the piece to move at least 10 spaces?

5. Joey is taller than Greg, who is taller than Rick, who is taller than Mike. How many different ways can they stand in line so that the tallest person is always last?

6. How many different combinations of quarters, nickels, dimes, and pennies can be used to make $0.25?

7. Roll a number cube 10 times and record the results. Repeat 3 times. Using your results, is there any way to predict which number the number cube will land?

	Roll 1	Roll 2	Roll 3	Roll 4	Roll 5	Roll 6	Roll 7	Roll 8	Roll 9	Roll 10
Set 1										
Set 2										
Set 3										

12-5 Skills Practice

Using Sampling to Predict

Describe each sample.

1. To evaluate the defect rate of its memory chips, an integrated circuit manufacturer tests every 100th chip off the production line.

2. Students who wish to represent the school at a school board meeting are asked to stop by the office after lunch.

3. To determine if the class understood the homework assignment, the math teacher checks the top 3 papers in the pile of collected homework.

4. To determine the representatives to the recess activities meeting, 2 students are selected at random from each homeroom.

5. A member of the cafeteria staff asks every fifth student leaving the cafeteria to rank 5 vegetables from most favorite to least favorite.

6. One bead for every member of the school orchestra is placed in a bag. All but 2 of the beads are white. Each member draws a bead from the bag, and the members who pick the non-white beads will represent the orchestra.

7. A real estate agent surveys people about their housing preferences at an open house for a luxury townhouse.

8. To determine the most popular children's programs, a television station asks parents to call in and complete a phone survey.

9. Two teachers from each school in the district are chosen at random to fill out a survey on classroom behavior.

10. Airline boarding passes are marked with red stars at random to decide which passengers should have their carry-on luggage inspected.

11. To determine how often people eat out, every tenth person entering a Chinese restaurant is surveyed.